Numerical Control
for
Machine Tools

Charles H. Barron
Assistant Professor
Applied Science and Industrial Technology
College of Continuing Education
Rochester Institute of Technology

McGraw-Hill Book Company
New York   St. Louis   San Francisco   Düsseldorf
Johannesburg   Kuala Lumpur   London   Mexico
Montreal   New Delhi   Panama   Rio de Janeiro
Singapore   Sydney   Toronto

Numerical Control
for
Machine Tools

Library of Congress Catalog Card Number 73-150456

-003824-4

89101112131415 MAMM 987654

This book was set in Lino number 21 by The Maple Press Company and printed on permanent paper and bound by Vail-Ballou Press, Inc. The designer was Barbara Ellwood;
the drawings were done by John Cordes, J. & R. Technical Services, Inc. The editors were Cary F. Baker, Jr., and Anne Marie Horowitz. Annette Wentz supervised production.

Contents

Preface

This text has been written to bring to the classroom and shop a comprehensive single source of information about the numerical control of machine tools.

True, many standards have already been set by machine tool builders and the controls industries. However, it is very evident to educators that this information must be pulled together and presented to beginners in numerical control in such a fashion as will promote understanding and attainment of skills in this area.

The people who can best profit from this text are tool designers, technical students, apprentices, machinists, process planners, maintenance men, and draftsmen. Some management executives may also benefit.

I have tried to develop a text that will encompass the hundreds of different N/C units manufactured here and abroad and that will provide a broad understanding of the areas of learning common to *all* N/C systems.

It will become evident as the reader progresses in this text that I have very definite ideas about what is important to an understanding of N/C. The best-informed individual is the one who is aware that important events had a beginning in history. Understanding this concept makes it easier to understand the present, and, in many cases, to forecast future developments.

I wish to thank the many machine tool and controls manufacturers who provided manuals, photos, and other assistance.

Also, I should like to acknowledge the patience, useful ideas, and encouragement of my many friends at Edison Technical and Industrial High School in Rochester, New York.

A special note of thanks to my colleague, Robert N. Klafehn, Assistant Professor in Engineering Graphics at

Rochester Institute of Technology for the drawings used in the text.

The groups of students at Rochester Institute of Technology who evaluated the material in this book and made many fine suggestions deserve my sincere appreciation. This applies especially to Mr. Mech, Mr. Sheehan, and Mr. Zierle of the Eastman Kodak Company for their extracurricular activities in my behalf.

To my wife, Arline, I wish to express heartfelt appreciation for the many hours of typing, her unbounded patience, and her ability to decipher my notes.

CHARLES H. BARRON

Introduction

Industrialists are taking a long, hard, and approving look at the concept of machining by numbers. N/C is composed basically of two units. These are the machine tool itself and the machine control unit that directs the operations.

The potential cost-cutting factors and elimination of operator errors are, in themselves, worthwhile reasons for investigating N/C.

Numerical control of machine tools was originally developed to solve the problems inherent in machining increasingly complex parts and tools for modern aircraft and space vehicles.

The world is in an era of technological change and improvement that will not be stopped. Modern executives and intelligent union leaders are already planning, if not implementing, ideas for the future. The status quo in technology is a phrase that means nothing, because even as this is written, new concepts have been formed.

There will, of course, be many refinements in numerical control, but the principles of removing metal efficiently and accurately must still be followed. A good background in machining, with knowledge of feeds, speeds, and tool design, combined with knowledge of properties of metals, plastics, and the newer exotic space-age materials are essential to a well-rounded expert for N/C. A solid mathematics background, with emphasis on basic arithmetic, algebra, trigonometry, and geometry, and strict attention to detail are also of major importance.

Anyone with average intelligence can be taught to program simple positioning or contouring of profiles, but this ability is worthless if the other criteria are not met. A truly competent N/C expert must have the attributes stated previously.

To summarize:

1 Numerical control is here to stay, and grow and change.
2 Anyone with expert knowledge of the machining field can be an integral part of this system.
3 Industrialists must keep alert for every change, and make numerical control work for them wherever practical.
4 The principles of removing metal must still be adhered to.
5 The machine tool itself is basically the same. Only the machine control unit directs its operations. The control unit does nothing unless told (programmed) to perform some task.

1
Background and Evolution

Where did it all start? To really understand what is happening today, it is necessary to go back in time to ferret out the beginnings of the concepts of numbers and computation. There will be no attempt to go into this in any great depth. Only some of the notable and pertinent advances as they occurred through the ages will be touched upon.

NUMBERS AND COUNTING

Imagine, if you will, a world just starting to move forward to its ultimate destiny. Go backward in time to the Stone Age. A dark, damp cave—cold and forbidding. Two or three humans are huddled there. They each have but one thought. Their stomachs are empty and growling with hunger. A couple of animals go past the mouth of the cave. One man looks out. Trying to convey information to his cavemates, he may point to his mouth to indicate the presence of food. To let them know how many animals are outside, he probably grunts twice or raises his fists to denote 2. With these gestures the Stone Age man has set the stage for a long journey into the realm of numbers and computation.

At this point in history man did not have any great need for numbers or counting. His possessions were few, and mostly of a temporary nature. Before the end of the Ice Age, if he counted at all, it was probably not beyond 2 or 3.

When the ice retreated, however, many nomads settled down more or less permanently and began to farm. This all started about 10,000 years ago. Even then there was not much sophistication in their numbering systems. Some tribes used 2 for a base, others used 3. Many used fingers and toes as handy counters. This enabled them to count to 20. Evidence of 20-based number systems still exists. The French designation for the number 80 is *quatre-vingt*, or four-twenty.

A natural step from this crude method of counting was to use pebbles on the ground. Then shells were strung on reeds or hide. Immediately, the logical evolution to the efficient little computer called an "abacus" can be seen.

Naturally, the concept of numbers and counting developed a long time before recorded history. It is easy to imagine that man had some sense of numbers, at least to the extent of being able to recognize when one or more objects were taken from a group.

Simple methods of counting became a necessity so that tribes could know how many members they had and how many enemies existed. Man also had to tally sheep and flocks of other animals in some way. Scratches in dirt, notches in wood, and knots in thongs were some of their meth-

Fig. 1-1 Chinese abacus. (*The Bettman Archive, Inc.*)

ods for counting. Sounds, perhaps vocal, were developed to tally a flock of sheep or chickens.

As the years ground on it became vital to try to get some systematized counting procedures. Numbers had to be arranged in basic groups. The matching process employed determined the size of these groups. Man's fingers furnished a convenient radix; for example, the number eleven with the fingers of two hands as a base is derived from *ein lifon*, meaning one left over or one over ten. *Twe-tig*, or twenty, means two tens, and so on.

Two, three, and four have been used as bases by many primitive tribes. The surviving kin of Stone Age man in Australia, New Guinea, and Brazil do not have numbers beyond 2 or 3. Some tribes of South America use a base of 4 even today.

The quinary, or base 5, scale was used extensively. This, of course, seems logical; one hand equals five. One, two, three, four, hand, hand and one, hand and two, etc. The peasants in Germany used a quinary scale right up to the 1800s.

The reader may well have deduced by now that a base of 12 may also have been used. In prehistoric times, lunations logically led to the use of base 12, especially in measurements. The vestiges of this system are still with us—12 inches in a foot, the hours on a clock, the months in a year.

There is evidence that a base of 20 was in use by American Indians. This was probably before they took to wearing moccasins.

Before going any further, it would be a good idea to pause here and pull together some of the myriad facts that have been unearthed up to this point. It might also be a good idea to see if the goals of this section are being adhered to. Does the information given so far really relate to numerical control? I am sure it does. It does, first, because of what is to come later on in the text, and second, because numerical control is "machining by numbers." Therefore, something should be known of the origin of numbers.

To summarize:

1. Stone Age man made scratches on rock or sand to denote numbers.
2. After the Ice Age, as man turned to farming or commerce, he developed bases on which to count. Using fingers and/or toes he began to compute.
3. Pebbles on the ground, then shells strung on thongs were forerunners of the abacus.
4. Systematized number groups or bases began to be developed.
5. Twos, threes, fours, fives, tens, twelves, twenties, and other bases began to be used.

WRITTEN NUMBER SYSTEMS

So far the written number systems have not been mentioned. There has been reference to scratches, pebbles, and vocal grunts for denoting numbers, but by far the most interesting segment has not been touched on.

Simple grouping systems have a base, upon which is built progressive count. For example, as far back as **3400** B.C. the Egyptians used a simple grouping system. Based on the scale of 10, the symbols for 1 and a few powers of 10 are illustrated in Fig. 1-2. These symbols were used additively; each symbol was repeated as required. It was cus-

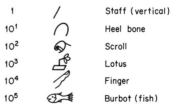

1	/	Staff (vertical)
10^1		Heel bone
10^2		Scroll
10^3		Lotus
10^4		Finger
10^5		Burbot (fish)

Fig. 1-2 Early Egyptian numbering symbols.

tomary, however, for the Egyptians to write from right to left, and not as shown in Fig. 1-3.

The Hindu-Arabic numeral system is the result of two sequences in the history of numbers. Some stone columns in India built around 250 B.C. are inscribed with some of our present number symbols. Some records cut in cave walls about 200 A.D. at Nasik are also specimens of our present numerals. One curious fact, however, is that there are no provisions for a zero, or any evidence of positional notation. Around 800 A.D. a zero and positional value were probably introduced in India. At any rate, the Persian mathematician al-Khowarizmi of Baghdad gives credence to a completed Hindu system at this time in history.

No one really knows when the new number symbols appeared in Europe or how they got there. Speculation is that this system was transported by commercial travelers and traders of the Mediterranean coast. Actually, there is a gap of about 200 years before the numbers appeared again in Spain.

The Arabs invaded the peninsula in 711 A.D. and remained for hundreds of years. During this period the Hindu numerals were reconstituted into a recognizable script known as Ghobar numerals. Ghobar is the Arabic word for sand or sandbox, which was then used for computations.

The 10-based positional notation system was eagerly adopted by European merchants. They found that they could do their bookkeeping much faster than with other systems. The bulky Roman numeral system was more cumbersome and much slower.

Fig. 1-3 The Egyptian symbols in this illustration are equal to 15,024.

Scientists, scholars, teachers, and the like were not so easy to convert at this time because of one serious drawback. The decimal numbering system had no way of denoting fractions. It was here that the ancient 60-based system was used in this vital area of computation. Some vestiges of the 60-based system remain today, as in degrees and designations for time. One thing that has been learned from history is that the world seems to be blessed with geniuses who arrive on the scene at just the right time. In the realm of numbers and counting, this fact also seems to be true.

The fifteenth-century director of the observatory at Samarkand, al-Kashi, perceived that negative powers could be used in the 10-based system as well as in the 60-based system. Events began to move quickly after this discovery. This is the case with most advances in man's development. At first there are long periods of strain to get moving, then advances proliferate, and finally, things move almost too fast to keep up with.

Further explanations of al-Kashi's treatise were produced in the sixteenth century by a German, Christoff Rudolff. Then a Belgian, Simon Stevin, produced a remarkable piece of work called "La Disme" (the art of tenths), a systematic document on decimal fractions. For thousands of years mankind had been searching for the fulfillment of mathematicians' efforts. Finally in 1617, a Scotsman, John Napier, gave us the little dot—that historic speck, the *decimal point*.

The irony of the whole search for fulfillment, however, is that now in the twentieth century our decimal counting system is not suitable for the computers in use today. Switches are either *on* or *off*. Consequently, only two numbers are used in even the most sophisticated computers, 0 and 1, off and on. The binary, or base 2, system of arithmetic may well be the answer for the future of mathematics. More of this interesting area later on in the text when it is discovered how computers work.

Fig. 1-4 Calculating machine of John Napier, Scotch mathematician, described in his book (1617). (*The Bettman Archive, Inc.*)

Since the last summary these facts have been unearthed:

1. Written symbols with base 10 appeared as early as 3400 B.C.
2. For hundreds of years there was no way of computing fractions in our chosen system.
3. It was only in 1617 A.D. that man finally got a decimal point to use as it is used today.
4. After the whole frustrating search, a return to binary notation for computers in this space age is noted.

This journey through time has been long, though fruitful. It is hoped that a better, deeper understanding of the evolution of numbers will enable the reader to more easily comprehend the intricacies of numerical control (N/C) as they apply today. Let us now change direction and trace some of the mechanical marvels as they evolved, concurrent with numbers and counting.

EVOLUTION OF COMPUTING MACHINES AND N/C

Shortly after John Napier's skillful victory over decimal fractions, a youthful French philosopher, Blaise Pascal, grew tired of laboriously adding and subtracting figures by hand for his father. In 1621 tax collecting and recording was a tedious and time-consuming job. The senior Pascal did the collecting, and young Pascal did the bookkeeping. This lad was only 19-years-old when he invented an adding and subtracting machine. This box was composed of cylinders and gears, a weighty accomplishment in itself for those times. From then on his job became easier, but still not to his personal liking. Pascal's device was very much like the desk-top calculators in use today.

Numbers from 0 to 9 were engraved on wheels. The wheel on the right stored integers 0 to 9. The second wheel stored the tens, the third wheel stored the hundreds, and so on. To store 285, for example, involved putting a 2 on the third wheel, an 8 on the second, and a 5 on the first. Adding 7 and 4 involved storing the 7 on the first wheel, and then turning the wheel 4 places. A series of gears then turned the next wheel.

In 1694 a machine that could multiply as well as add was built to plans designed by Liebnitz. This calculator turned out to be very unreliable.

As early as 1728 knitting machines were automated to produce varied patterns by the use of holes punched in steel cards. A little of that story is very relevant to the goals of this chapter.

Weaving at this time was a fiercely competitive business. A Frenchman, Joseph M. Jacquard, saw the need for a better way of weaving patterns in materials than the slow and expen-

Fig. 1-5 Calculating machine invented by Blaise Pascal (1623–1662), French scientist and philosopher. (*The Bettman Archive, Inc.*)

sive processes in use at the time. By punching holes in steel cards and arranging them on his looms in varied ways, he was able to change patterns more quickly. First and foremost was the fact that he could do this with little cost. The presence or absence of a hole dictated whether or not a needle would be activated. This method became the forerunner of the cards and computers in use today. A patent was granted in 1801, and soon there appeared thousands of looms doing precise and reasonably priced work. Jacquard was very proud of his automatic loom, and actually used these cards to weave a portrait of himself in silk. Over 20,000 cards were used, but the artwork and the uniqueness of it was phenomenal for the times.

Many men deserve credit for their contributions to the field of mathematics and to the evolution of mechanical or other devices for counting. One of the giants was a man by the name of Charles Babbage. A banker's son, he was born in England in 1792. A large fortune was left to him that he used for his first love—mathematical and scientific

Fig. 1-6 The Jacquard loom (Engraving-1877). (*The Bettman Archive, Inc.*)

Fig. 1-7 Charles Babbage (1792–1871), English mathematician, perfected calculating machine, "Babbage's Folly." (*The Bettman Archive, Inc.*)

experimentation. He was self-taught, and evidently a mathematics genius, for at the age of 18 when he went to Cambridge, they could teach him nothing about algebra that he did not already know. Babbage's story is a lengthy one, so only facts that are germane to our text will be dwelled on here. In 1822 he completed a "difference engine" for computation of polynomials. This machine was essentially an adding machine and worked reasonably well for calculating tables to an accuracy of six places. Charles Babbage was not easily satisfied, however, and so he next proposed a machine capable of accuracy to 20 decimal places. The British Government respected this wizard's knowledge and ability, and paid out about 17,000 pounds for its construction. For 4 years Babbage and his coworkers toiled toward this end. His ideas and designs expanded the size and complexity of this invention until finally he was forced to machine some parts himself because of their close tolerances and complex geometry. He decided to scrap what he was doing and proceed to a machine that could do not only all the arithmetical calculations, but could also control itself and use punched cards such as Jacquard had used in his looms.

Babbage's fertile mind and the ideas it spewed forth were far too advanced for the technology of that time. His "analytical engine" was to have all the functions that today's computers have, including storage, mill, memory, input, and output. It was never finished, but the man would have loved dearly to see today's modern computers doing all that he had so fervently predicted they could do. Charles Babbage died in 1871 after spending 40 years of his life trying to perfect his analytical engine. His work was forgotten until the early 1900s when others began to dream as he had. Their dreams, of course, culminated in what are known today as electronic computers.

As is the case when research tends to become astronomical in cost, the United States government can usually be depended on to participate, especially if the interests and/or

Fig. 1-8 Part of differential calculating machine on which Charles Babbage, English mathematician, worked for 37 years, only to find it rejected for patents by the government. (*The Bettman Archive, Inc.*)

Fig. 1-9 Early calculating machine. Card is removed from the press and deposited in the sorting box. Census taker (*Hollerith*). (*The Bettman Archive, Inc.*)

defense of the nation are involved. The United States Air Force became aware by 1949 that the highly complex shapes that were needed for space-age aircraft and missiles were causing us to lag dangerously behind in some projected target dates.

The exotic metals and other materials were also causing some difficulties in fabrication. Evidence that it was no longer possible to continue with conventional methods of machining began to pile up.

John Parsons, a pioneer in numerical control, procured a contract from the Air Material Command "to study the design problems involved in making an experimental machine that would produce contoured surfaces from instructions in the form of punched cards (I.B.M. standards)." He designed and built a prototype of a planer mill that machined various contours and pockets.

The Massachusetts Institute of Technology agreed to help in refining the basic machine, and to attempt to incorporate self-checking features as part of the whole numerical control package. The contract with MIT had these features:

1. Develop coding for punched tape to reduce the number of tapes required
2. Provide checking features to minimize error in the event of equipment failure or malfunction
3. Design and develop equipment to read the tape; convert the instructions to electronic signals; and use these signals to control movements of table or cutter in longitudinal, cross, and vertical travel simultaneously
4. Modify an existing tracer-controlled milling machine to test the system and point the way for further development and improvement
5. Demonstrate that it would be technically feasible to build and use machine tools employing punched-tape control

This last requirement was the one feature that the Air Force was actually waiting for. If this could be proven, then America was back in the race.

In 1950 a prototype one-axis machine was in operation. Advances piled one on top of another, and by 1952 a three-axis machine was operating in the laboratory at MIT. In that year 23 companies that were engaged in government contract work availed themselves of MIT's efforts and proceeded to overcome former time and cost factors with numerical control of machine tools.

Fig. 1-10 N/C machining center. (*Giddings & Lewis Machine Tool Co.*)

It is well to remember that these first machines were highly sophisticated contouring or continuous-path machines capable of machining complex arcs, angles, pockets, etc.

The next development was a simpler point-to-point, or positional control for drills, milling machines, and even lathes. The cost factor for these first N/C machine tools dropped accordingly and eliminated much of the need for computer time for programming.

As will be explained later in the text, manual (written) programming for point-to-point N/C is comparatively simple. The future is for those who understand and can use this new concept.

From the discovery of numbers and counting to the complex methods for removing metal in use today has been a long journey for man. Numerical control is here to stay. The first sentence in this chapter asked, "Where did it all start?" Now the question is, "Where do we go from here?"

THE CONCEPT OF NUMERICAL CONTROL

It has been pointed out that specific treatment of a certain control or machine tool is not the goal of this text. Certainly some examples will be given to clarify information, but only after all the principles and facts that are common to all numerical control systems are understood.

It was also said earlier that N/C is the operation of machine tools by numbers. This statement is valid, but there are a few details that should be cleared up. What numbers? How are they presented? How does one begin? What exactly is meant? For the learner who, at the moment, has never seen N/C at work, these are but a few of the questions that are unavoidable. Be assured that these questions and many others will be answered, but in the correct sequence.

First it is necessary to understand what the N/C process is. Formerly, the machine-tool operator guided a cutting tool around a workpiece by manipulating handwheels and dials to get a finished or somewhat finished part. In his procedure many judgments of speeds, feeds, mathematics, and sometimes even tool configuration were his responsibility. The number of judgments the machinist had to make usually depended on the type of shop he worked in and the kind of organization that prevailed. If his judgment was in error, it resulted in rejects, or at best, parts to be reworked or repaired in some fashion.

Decisions concerning the efficient and correct use of the machine tool, then, depended upon the craftsmanship, knowledge, and skill of the machinist himself. It is rare, indeed, when even two expert operators produce identical parts using identical procedures and identical judgments of speeds, feeds, and tooling. In fact, even one craftsman may not proceed in the same manner the second time around. This matter of skill and craftsmanship is a serious one for the manufacturer. Recent surveys by the Federal government, individual states, and local communities bear out the distressing fact that America is not in very good shape in the skilled trade areas. You have only to read your newspapers to see how training programs are springing up all over your own city or county. These may be fine programs, but staffing them with experienced instructors and getting trainees creates even more problems.

Numerical control will not solve all of our problems, but it does alleviate some troublesome areas. It is definitely one answer. With N/C the correct and most efficient use of a machine no longer rests with the operator. Judgments and procedures are not his province. Process planners and programmers, sometimes far removed from the shop, now have the responsibilities for these matters.

What then is the function of the machine operator when N/C is installed? One fact is sure. Not all workpieces require machining with N/C. This decision of whether or not to route jobs to an N/C machine or to a man in the shop is determined many times by cost factors and availability of machines. The operator of a numerically controlled machine monitors the operation. In many cases he does not need to be a skilled machinist. Familiarity with programs and tapes, or cards, plus being able to observe intelligently and make suggestions to improve a program are all good steps in the right direction. Intelligence and adaptability are essential.

A good attitude toward a man's job seems to be a by-product of N/C development. Workers are proud to be so closely associated with this new development, and generally become very good operators, even if previously they had been evaluated as only average. They sense that the opportunities for advancement in this field are unlimited.

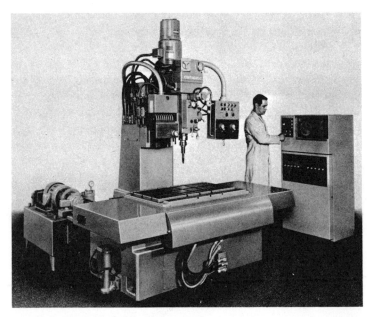

Fig. 1-11 Vertical Cintimatic machining center. (*Cincinnati Lathe &
Tool Co.*)

To understand what N/C is, then, two things should
be made clear. First, the actual machine tool with a capable
operator can do nothing more than it was capable of doing
before a machine control unit (MCU) was joined to it. New
metal-removing principles are not involved. Cutters cut as
they always have. The same drills, milling cutters, or other
tools are still needed in N/C as they are in traditional machining. Cutting speeds, feeds, and tooling principles must
still be adhered to. What then is the principal advantage?
For one thing, the idle time, or time to move into position
for new cuts, is limited only by the machine's capacity to
respond. The utilization rate (chip making) is, therefore,

Fig. 1-12 Precision jig borer, 2E with Pratt & Whitney Control. (*Colt Industries, Inc., Pratt & Whitney, Inc., Machine Tool division.*)

much higher than on a manually operated machine. MCU can deliver rapid and precise instructions to the machine. When N/C machine tools are observed, notice how ruggedly they are built. The actual utilization rate, being higher, now compresses into 1 or 2 years the use that a conventional machine receives in possibly 10 years.

The second characteristic of numerical control that should be understood is that the N/C unit is a slave. It can initiate nothing. It cannot think. It is not a mechanical brain. It cannot judge and is not able to adapt to changes by itself. Without directions in the form of punched tape or cards, it will stand until it crumbles—just a beautiful lump of steel, wires, cables, and paint. Once the operator places the program medium (tape) in the MCU, loads the workpiece on the machine, and presses the button to initate the cycle of operations, then, and only then, does the N/C machine come alive.

Even then on some machines the operator may have to perform some functions that are not a part of the program. Some N/C machine tools position the work under the cutter, but the operator must feed the tool into the workpiece manually. The more sophisticated units are completely tape-controlled, and leave only the loading and unloading operations to the operator. Aside from the advantages already discussed, there are others that relate directly to the whole concept of N/C.

The reader should have a good grasp of these other compelling reasons for the growth of N/C, if only to get organized in his own mind the real worth of this exciting development. Consider the following:

1. The elimination or reduction of costly setups. Simple clamps are taking the place of many jigs and fixtures; the program tape positions the work or cutting tool in the right place.
2. The design, manufacture, and storage of jigs and fixtures is practically eliminated.
3. Complex machining operations are more easily accomplished with tape.
4. On some N/C units, standard single-point cutting tools eliminate the need for form tools that are costly to produce and change if the design of parts is changed.
5. If setup time and operator calculations are reduced, the utilization rate soars correspondingly. One N/C machine may do the work of two or more conventional machines.
6. Lead time, or the time needed before a job can be put on a machine due to the need for special tools such as templates, is obviously reduced.

Fig. 1-13 Horizontal Cintimatic machining center. (*Cincinnati Lathe & Tool Co.*)

7. Families of parts, those that are pretty much alike with only slight variations, lend themselves to N/C very well. Modifying a program is faster and cheaper than having to make new tools and fixtures for each kind of workpiece.

8. Elimination of human error is also a by-product of a program that has been checked out. The program will work without error each time it is used. Only a machine or control malfunction can interfere with the activities of a good program.

9. Storing of tapes is far less costly than storing parts and fixtures. This item is a real saver.

10. In many cases parts are completely machined on one N/C machining center. It is costly and time-consuming to wait for machine time in various shop departments before shipment. Routing through departments for drilling, milling, burring, etc., can be eliminated in many instances.

11. Managers and executives have found that once they have installed N/C they have learned some startling facts about operations done in the conventional manner. Idle time, or time when a tool is not making chips, has sometimes run as high as 90 percent of the total time used to complete a workpiece on manually operated machine tools.

12. Machining costs are predictable with great accuracy.

13. Precise hole patterns are attainable.

14. Operator fatigue is now not a factor. An N/C machine may run three shifts per day. It does not get tired.

15. Many complex workpieces are now machined from the solid instead of from castings that require expensive tooling.

From the foregoing it appears that the concept of numerically controlled machine tools is a good solid advance in machining technology, and one that shows evidence of becoming even more advantageous in the future.

REVIEW QUESTIONS

1-1. Did man have a need for numbers and counting in his earliest stages of development? Why?

1-2. As man began to settle down in groups and farm the land, did a need arise for some numbering and counting systems? State some reasons for your answer.

1-3. Can you think of a way by which a man might have counted his flock before symbols in writing were developed, other than by the methods described in this chapter? Illustrate.

1-4. In positional notation it is important to place numbers in their exact locations if the value is to be interpreted correctly. Does the same rule apply in an additive numbering system? Show by examples the innate differences in each system, if any.

1-5. Why is a 10-based positional numbering system superior to the Roman numbering system, especially for commerce? Illustrate.

1-6. Is the abacus a kind of computer? Give some reasons for your answer.

1-7. Why did the United States Air Force decide to help finance research in numerical control of machine tools?

1-8. John Parsons is often alluded to as the father of N/C. What educational institution helped in the refinements of N/C? In what year was the first numerically controlled three-axis machine in operation?

1-9. Were the first N/C machines simple drilling machines? Elaborate.

1-10. Since cost was a factor that helped determine whether N/C machines were to be purchased by industry, what was the next logical development (after early contouring machines) in the manufacture of numerically controlled machine tools?

1-11. In conventional machining, what are some of the judgments that a machinist may personally have to make when producing a workpiece to blueprint dimensions?

1-12. Why is it necessary now to "put the skill into the machines?"

1-13. Does the operator of an N/C machine have to be a skilled machinist?

1-14. Are new principles of material removal involved in N/C machining? Explain your answer.

1-15. Is the chip-making rate higher or lower for an N/C machine as opposed to its conventional counterpart for a given workpiece? Why?

1-16. If the design, manufacture, and storage of jigs and fixtures is practically eliminated, what, then, is stored?

1-17. Why can complex machining operations be accomplished more easily with tape?

1-18. Why do families of parts lend themselves so well to N/C?

1-19. What are some of the advantages of machining centers?

1-20. Aside from the economic advantages of N/C, what are some of the operator advantages?

2
Numerical Controls—How They Operate

A world-wide shortage of skilled craftsmen has been shown to be a major contributing factor in the development of numerical control systems. By putting the skills into the machine and its controller or director, many problems have been alleviated.

In order to accomplish the above, it was imperative that the machine control unit perform many of the functions of the skilled machinist. Machining, or the removal of metal or other materials to produce a product that is made to specifications on a drawing or blueprint, requires many skillful judgments and other considerations when performed by individuals.

To understand what a machine control unit must do, and how it does these things, it might be well to review some functions that the machinist performs.

1. Studies the blueprint or sketch, or, in some cases, a sample workpiece
2. Plans a sequence of operations by writing or mentally preparing the sequence
3. Does calculations for speeds, feeds, and dimensions when necessary
4. Selects materials, and the machine tools and cutters to be used

HOW THE MACHINIST PRODUCES THE PART

When the craftsman has prepared his roadmap, before actually machining anything he has, in essence, created a manuscript with instructions telling him what must be done. *How* he machines the workpiece is a product of his experiences in manipulations of dials, measuring instruments, cutting tools, and numerous other pieces of specialized equipment and supplies. In any case, he knows the exact relationship of the work to the tool's cutting edges, so that by manually operating dials or other devices he can be certain that the final results will be accurate as material is removed. In the above functions a craftsman has collected, stored, and transmitted information, which is essentially what a control system does.

HOW NUMERICAL CONTROLS COLLECT INFORMATION

When a manuscript has been prepared and the tape has been punched for a specific job, the stage has been set for the controller and the machine tool. The coded information on the tape then has to be given to the director by some means. In numerical control a device called a reader is used to send

Fig. 2-1 Continuous-path control showing photoelectric reader in lower window. Also shown is the manual data input section, four tool offset boards, and feed-rate overide. (*General Electric Company*)

this information to the control unit. Readers may be mechanical, photoelectric, or, in some applications, even pneumatic. They may read a single row of information at a time, or they may read a complete block and transmit information to the control all at once, instead of row by row. Reader speeds vary considerably. Most mechanical readers are usually capable of reading 60 characters per second, while others are even faster. Photoelectric readers are available with reading speeds to 1,000·characters per second.

ELECTROMECHANICAL READERS
Mechanical tape readers are used mostly on point-to-point N/C systems where little or no contouring is done. This

Fig. 2-2 Electromechanical tape verifier. (*Friden Inc., division of the Singer Company*)

is not to say that they may not be used for contouring. They are reliable, easy to maintain, and the cost is relatively low. The slower reading speeds are necessarily a drawback in this kind of application, however, especially when hundreds of rows of information are needed to complete a large radius or other configuration. These readers mechanically sense the codes punched into tape and convert each code into electrical impulses that are transmitted into the control unit. They sense the codes with pins that, when they enter a hole present in the tape, make contact and create the respective impulse. The sprocket on the reader feeds the tape past these pins at a predetermined rate of speed. The sensing pins are always in contact with the tape as it advances.

PHOTOELECTRIC READERS

These readers are excellent for use on contouring controls. Light shines through any holes in the tape. Beams that pass through the holes to a photocell are converted to electrical energy, and amplifiers strengthen the signal enough to make it adequate for use in the controller. This type of reader has no drive sprockets to drive the tape. Capstans, either rotating with a brake system for stopping the tape or direct-drive capstans with printed circuit motors, are used. Braking and acceleration are almost instantaneous. Reading speeds up to 1,000 characters per second make smooth motions on arcs and other configurations possible.

AFTER THE TAPE HAS BEEN READ

When the reader transmits the coded information to the control in the form of signals, the part of the controller that accepts information now performs its functions. Registers accept proper coordinate information for the various axes, preparatory commands, and auxiliary and miscellaneous functions. This information goes to the respective registers or sections where actuation signals are relayed to the machine tool drives.

Some controls are also equipped with what is known as a buffer storage section. This allows the control to accept information in the buffer register while an operation is being done at the machine from the active storage register. When that operation is completed there is an immediate transfer from buffer storage to the machine actuation registers. This helps to lessen the time between reading and performance. In fact, the transfer to the active register from buffer storage can be considered instantaneous, providing a smooth motion around curves, angles, or other configurations.

After the commands have been signaled to the machine tool, and slides or spindles start to position, there may or may not be information fed back to the control that the machine elements have indeed gone to the precise locations as commanded by the control. This would depend on whether the control used is an open-loop or a closed-loop system.

CLOSED-LOOP AND OPEN-LOOP SYSTEMS

Numerical control systems must perform many of the functions of the individual craftsman. In order to do this the controller may have information fed back to it. A machinist who actually turns handwheels and reads dials or uses other devices can check visually or with various measuring instruments to ascertain, for example, whether the center of a spindle has been accurately positioned. These devices feed back to the craftsman information that tells him whether an element is or is not in location. If it is not, then he can physically adjust the elements involved to gain positional accuracy.

A numerical control system that has this kind of feedback of information can compensate for errors in positioning when necessary. This is called a closed-loop system. Many N/C installations are of this type. They utilize measuring devices on the moving slides or flat surfaces of the machine tool to record exact position. If there should be a discrepancy between the commanded position and the actual location, then automatic compensation occurs to rectify the error.

The measuring devices used for this purpose are many and varied. The kind that is used is very important to the purchaser of N/C equipment, since accuracy and dependability are directly affected by the feed-back system.

Open-loop systems provide no check or assurance that a commanded position has actually been achieved. There is no feedback of information to the control. For applications that do not require extremely accurate positioning, these systems are quite excellent. In fact, they are being perfected and used more frequently as more is learned about N/C. Systems that depend mainly on the accuracy of the machine elements gained wide acceptance outside the United States, more recently in this country. Much of the cost of the N/C package is involved in the closed-loop sections of the control and machine.

ELEMENTS OF A CLOSED-LOOP SYSTEM

The closed-loop or servomechanism type of system has basically these elements:

1. A variable to be controlled
2. A device that constantly measures or monitors the variable, and signals the control as to the condition or status (transducer)
3. A comparator that receives the signal and compares the actual status with the desired conditions, detects any discrepancy, and signals this error to the control
4. A compensation command from the controller that reduces the error to zero, and the compensation command that becomes zero

If all of these elements are present, then it is truly a closed loop. The variable to be controlled may be any

process from controlling temperature in a freezer or house
to positioning a cutter or slide on a machine tool with N/C.

Two examples of commonly understood feed-back systems are heating plants that are thermostatically controlled,
and the operation of an automobile. When a person drives
a car, he or she visually checks the speedometer. The actual
speed is compared with the desired speed, which may be
posted as a limit. The human control or brain notes the
difference between the actual speed and the posted speed, and
commands the foot to adjust the condition until the correct
speed is attained. When this set of conditions is secured,
the foot stops motion until another error signal, if any, is
fed back to the driver. See Fig. 2-3.

Feed-back devices The closed-loop or servomechanism type
of system selected for an N/C installation requires careful
evaluation by the purchaser. It must provide dependability
and accuracy. A less than adequate feed-back system can
result in work being scrapped, even when machined on very
expensive and sophisticated N/C equipment. Many kinds

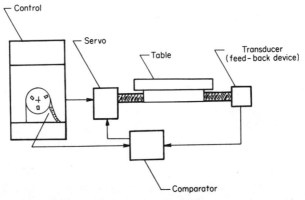

Fig. 2-3 Diagram of a closed-loop system.

Fig. 2-4 Feed-back device. Light beam behind disk is interrupted as disk revolves on lead screw. This counting device is very accurate at high speeds. This quantizer disk was photographed on a Pratt & Whitney Tape-O-Matic drilling and milling machine located in the N/C lab at Edison Technical and Industrial High School, Rochester, New York.

Fig. 2-5 This illustration shows the feed-back resolver assembly on the ball screw of a tape-controlled, continuous-path lathe. (*Lodge & Shipley Co.*)

of feed-back devices are available. Electric scales, electro-optical scales, magnetic scales, synchros, shaft digitizers, resolvers, laser types, linear or rotary transducers, and many other types are used on numerical controls.

Just as there is much that is common to all numerical control systems, all of the aforementioned devices have characteristics that can be categorized. Feed-back devices may be:

1. Digital or analog
2. Absolute or incremental
3. Subject to wear or not subject to wear

Digital devices   Digital systems count down linear or rotary motion in minimum increments on machine lead screws or other members, such as racks and pinions. Pulses are generated and fed back to the control. When the countdown in the control registers zero from the taped command, the machine element will stop. It should be noted that a slowdown of the moving element occurs at a predetermined amount from countdown zero so that overshoot does not occur.

Analog feed-back devices   Analog devices sense and constantly monitor variations in levels of voltage. The moving member on a machine may overshoot in both directions and hunt for the position that allows it to stop. This position is achieved when the voltage null point is reached. Nulling occurs when the actual signal strength is equivalent to the reference signal strength.

Absolute and incremental devices   A feed-back device that is absolute always knows its actual position. Each position has a specific value. These devices are highly immune to electrical errors or power failures. Position readouts can be depended upon to be accurate, and are recommended for use in applications where scrap cannot be tolerated. Incremental feed-back devices, while less expensive than absolute systems, are susceptible to errors that can be caused by false electrical signals. Absolute systems can correct a reading error on the next reading. Incremental systems cannot make these adjustments. They are, however, quite excellent for use on N/C systems such as drills, and where the value of the workpieces is not high.

IMMUNITY TO WEAR   Feed-back devices for measuring position may be immune to wear. There is no physical connection between the moving machine element and the measuring device. Position is measured directly. Electro-optical scales and similar devices fall into this category.

Rack and pinion measuring devices, and others of a mechanical nature, are subject to wear. The quality of most of these, however, is very high, and wear-resistance is excellent. It is suggested that the instructor and students procure machine and control manuals from manufacturers and study the many types of feed-back systems that are available.

MACHINE-ELEMENT DRIVES FOR N/C

The words servo, servo motor, and servomechanism have their derivation from the Latin word *servus* (slave, servant). These drives are required in numerical control because of the many changes of direction and motion velocity needed on point-to-point, or continuous path, systems. Accuracy when applying feed motions, distances, and speeds is desirable and necessary. Direction changes and reversal must be smooth and reliable.

Servos may be electrical or hydraulic. Hydraulic drives are preferred if operations are at maximum horsepower, and fast, smooth response is desired. There are so many variables, however, that enter the picture, that the choice of electrical or hydraulic power servos should be evaluated in relation to the whole N/C installation.

SOME FEATURES AVAILABLE FOR N/C SYSTEMS

Any unfamiliar terms used here will be explained later in the text or Glossary.

In addition to zero-shift capabilities; tool offset; feed and speed override; sequence-number readout; dwell; and linear, circular, and parabolic interpolation, there are some controls that offer additional features as standard or as options. A brief explanation of these added features follows:

SEQUENCE NUMBER SEARCH

By manually dialing in the desired block number the operator can quickly find any sequence number he desires. The reader will feed the tape to the sequence number asked for, and display the number in the sequence readout window. Another variation of this is the feature that searches out blocks that are identified by a letter other than N. These are complete blocks of information, and each time the operator depresses the search button, the reader will stop on one of these blocks.

COMMAND READOUT

Command readout is standard, or should be, on all continuous-path systems. It is very helpful when tape codes are displayed as they are read. This feature is very helpful when corrections have to be made in a program, or for reorientation of the machine when tool adjustment or replacement is necessary.

POSITION READOUT

Knowing where a tool tip or machine slide is at anytime in a program can be very helpful. This feature is not usually standard, and the type of feed-back system determines whether it is practical or not.

PECK DRILL

This mode is useful when holes are to be drilled that may be too deep for one pass. The drilling operation may be programmed to drill for a short distance, retract to clear chips, and provide for more coolant in the hole for subsequent passes until full depth is attained.

PARTIAL RETRACT

This feature provides for tool movement to new locations without full retraction. Tool retract to gage height saves much time when many locations are to be machined by a specific tool.   Drills and milling machines may have this feature as an option.

TOOL-LENGTH COMPENSATION

Differences in tool lengths are compensated for by this feature.   The operator manually adjusts for the differences in lengths of the tools by use of switches or dials at the control before a program starts.

MANUAL DATA INPUT

A complete block of information, with the exception of sequence number, may be programmed manually at the control by the operator.   Tape programs can be stopped for tool inspection by manually inserting data to back off, and then manually programming a return to position.   It is also a valuable feature when checking out machine and control operation for maintenance purposes.

New features are becoming available at a rapid rate. Experiences with numerical control seem to create more and more demands by users for options that facilitate programming and operation.

TRENDS

There is no question that N/C is going to influence much of the future development of manufacturing.   People will also be affected by the advanced techniques that will be incorporated into numerical control systems.   If planning is not done now to soften the impact of the probable reduced need for unskilled machine operators, then the effects can be considered to be less than desirable.   N/C machines are helping to build N/C machines.   Minimal skills and fewer people may be all that is needed to operate and maintain N/C installations in the future.   As more reliability and lower costs are developed, it is entirely believable that systems may be allowed to operate unattended.   Machining centers that finish all sides of workpieces without human handling are examples of this concept.

As with most technological advances, the size of control packages will decrease due to plug-in modular circuits and the application of microcircuits that have been developed already to the point that a complete circuit is difficult to see with even a jeweler's loupe.

Tapes and tape readers will gradually be phased out as interfaces connected directly from computers bypass these items. This concept of a remote computer directing machine tools is already a reality. Another variation of this is a direct computer-to-machine hookup to eliminate cables and other hardware leading to remote computers. This type of installation is also available now. The computer in this case needs only the capability to direct the specific machine. All other extraneous capabilities of standard computers need not be part of the package.

Machine tools are now being designed specifically for N/C application. As long as the machine operator no longer must manipulate handwheels and levers and be on top of each operation, the basic machine tools are becoming barely recognizable as simply lathes, milling machines, grinders, et al. Easier chip removal and nontraditional location of cutters are part of the new design considerations.

Adaptive controls that sense all the forces of the chip-making processes and feed information back to the control so that it can adapt to specific situations are already in existence. One application of the use of adaptive controls is in a case where a cutter may start to dull. Measurement of the forces involved may, when fed back to the control, cause modifications to be made in speeds and feeds.

Many of the trends and realities that have been discussed here are by their very nature too expensive for small shop owners to consider. They must, however, be sure to know what is new and anticipated. The initial cost of installing N/C in a shop should not be a deterrent all by itself.

There are many other considerations that, when evaluated carefully, may cause a prospective user to reconsider.

Trying to anticipate the future of numerical control is a study in itself. It is certain, however, that the surface has barely been scratched. Any application that requires motions to be controlled is sure to be scrutinized carefully to see if numerical control can be applied.

REVIEW QUESTIONS

2-1. When a craftsman produces a finished workpiece, what has he done that is analogous to what a numerical control system does?

2-2. What is the device called that reads the information on the tape for transmission to the controller?

2-3. What types of readers are available?

2-4. Do all readers read a single row of information at a time?

2-5. Do reader speeds vary? Elaborate.

2-6. Slow reading speeds are a drawback in what kinds of applications?

2-7. What type of reader would be best for contouring applications?

2-8. What are the advantages of buffer storage?

2-9. When no information is transmitted back to the control that the machine elements have indeed gone to the commanded locations, what is the system called?

2-10. In your own words describe what a closed-loop system does.

2-11. What are the advantages of absolute feed-back devices?

2-12. What type of mechanical feed-back device is used in some N/C applications?

2-13. Which type of servo is preferred for maximum horsepower and fast, smooth response?

2-14. What is peck drilling?

2-15. What is an important advantage of the partial retract feature?

2-16. As technological refinements occur in N/C, what are some of the predictable trends?

3
Classifications of Point-to-Point Systems

In Chap. 1 when evolution of numerically controlled machines was studied, you will remember that the first N/C units built were very complex, contouring types. Computers were used with these units, and it was thought at that time that a very highly educated engineer would have to do the programming. American machine builders were engulfed in a confusion of facts, figures, and false impressions. They soon found that a simpler control system would take care of a large segment of industry. In fact, sales of this simpler system far outnumber the sophisticated contouring controls purchased today.

There are two basic classifications of systems: positioning, or point-to-point, and continuous path, or contouring. Consider first the simpler positioning numerical control machine tools.

POINT-TO-POINT SYSTEMS

Positioning, or point-to-point, is simply the movement of the machine table (with the workpiece clamped or loaded in some manner) to a specific position under the cutting tool. On some point-to-point machines the spindle travels over the workpiece to a specific point. In either case, the path taken to get to a position is of no consequence since the cutter is withdrawn while traveling to each point. Some point-to-point machines are equipped with milling capabilities also. This will be explained later when a simple job is programmed. Many programs for positioning systems are written manually instead of by computer, as are some of the less complex programs for contouring or continuous-path machine tools.

WHAT THE PROGRAMMER HAS TO KNOW

The programmer studies the engineer's drawing and translates the operations to be performed onto a manuscript in a prescribed format. The program is then typed and a coded tape is the by-product of typing.

The individual who programs the job for N/C generally:

1. Studies the drawing
2. Chooses the N/C machine tool to be used
3. Identifies the type of material to be machined
4. Knows the format and functions of the machine he has chosen, which can be put on tape
5. Checks the tooling needed
6. Establishes the procedure
7. Calculates speed and feeds and other considerations
8. Prepares a manuscript for typing
9. Receives a tape, a by-product of the typing, as a master

to be kept on file (some plants destroy the duplicate as soon as a job has been completed). (See flow chart Fig. 3-1.)

In many cases a process planner establishes all the data on tools, procedures, speeds, feeds, etc., and the programmer simply prepares the manuscript. Some point-to-point machine tools have only simple movements in two axes under tape control—left, right and in, out.

Other positioning machines may be controlled on three or more axes and have tape-controlled spindle speeds, direction of spindle (clockwise or counterclockwise), feed rates, tool indexing, coolant on, coolant off, tool selections, and others.

Any of these functions not able to be controlled on tape due to the limitations of the N/C machine tool and controls, must then become a function performed by the machine operator. He should be provided with a complete program of instructions. The fewer judgments and decisions to be made at the machine by the operator will enhance the overall efficiency of the numerical control concept in the respective plant. The responsibility for every phase of the program from start to completion should rest upstream with the process planning department, after consultation with the design engineering department.

Before actually beginning to manually program a job, there are a few points to be thoroughly understood in order to have the learner become as proficient as he is capable of being. The first few problems may seem to have no bearing on N/C, but they *do*, and you will be a better student of this concept for having followed through. Learning to program parts for a specific machine tool made by a specific machine builder is not the goal of this text. The aim is to be able to adapt to any system without too much effort or loss of time.

Aside from the considerations of planning how the job

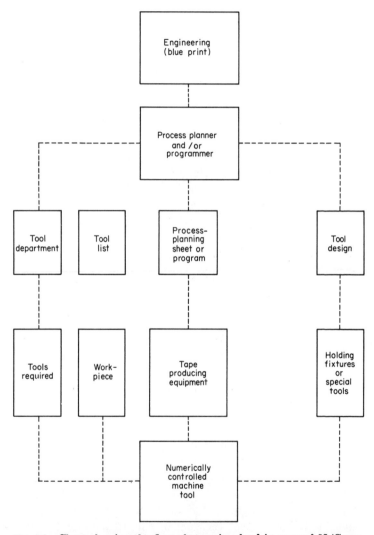

Fig. 3-1 Chart showing the flow of steps involved in manual N/C programming, from the drawing to the machine tool.

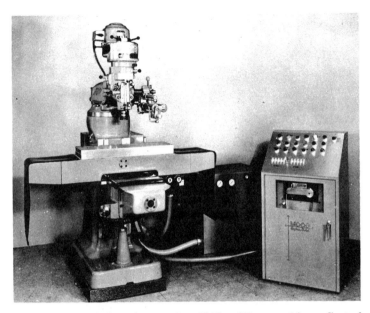

Fig. 3-2 A vertical, point-to-point N/C milling machine. Control accepts the fixed-block format. Canned cycles, coolant control, tool-change indicator light with buzzer from tape. (*Hydra-Point division, Moog Inc.*)

is to be done, there are certain things that have to be known about preparing a tape that is compatible with the N/C machine it is going to be used on. A tape prepared for one MCU may not even work on another kind of control. The programmer must know the following about any N/C machine he may be concerned with:

1. Is it a point-to-point or continuous-path system?
2. Is it an absolute system?   Is it incremental?
3. Does it have a fixed zero with a small amount of shift, full-range zero shift, or full floating zero?

4. Is the format tab sequential, fixed block, word address, or variable block?

There are other variables, of course, and these shall be covered in subsequent chapters.

You have learned what a point-to-point system is, so leave that question for the present. Is it an absolute or incremental system? It can be shown by a simple example of each that there is no great difficulty in determining which of these two a control system might be.

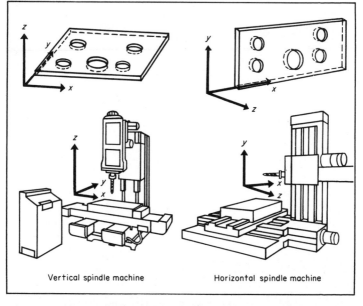

Vertical spindle machine   Horizontal spindle machine

Fig. 3-3   An illustration showing notations for x, y, and z axes on horizontal and vertical milling machines. (*Friden, Inc., division of the Singer Co.*)

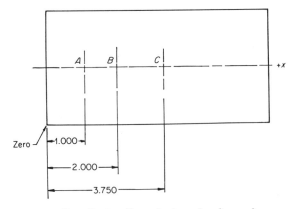

Fig. 3-4 Coordinate dimensioning showing references from datum zero.

ABSOLUTE SYSTEMS

Absolute systems position slides from a reference point. This means that *all* dimensions are calculated from one datum point or zero. Jig-borer-type dimensioning is excellent for an absolute system. In this example only the x axis will be used to demonstrate the absolute system. Study Figs. 3-3 to 3-5.

Intersection	Absolute dimension from zero reference
A	$x = 1$ inch
B	$x = 2$ inches
C	$x = 3\frac{3}{4}$ inches

Fig. 3-5 Notations for points A, B, and C in Fig. 3-4.

The table or tool does not have to return to zero before each move. Only the programmer must consider this in preparing his manuscript. A very important point has been made here in relation to drawings. Before the draftsman does anything else, he should establish his dimensions in such a way as to eliminate the possibilities for error by the programmer. He may, of course, establish the datum point according to the requirements of the job. Engineering and drafting personnel should have the responsibility for making drawings from the standpoint of what the numerical control system is best suited to. This is only one of the reasons why all personnel should learn everything they can about all N/C systems installed in a plant. The whole shop from the top down to the operator must gear themselves and their procedures to numerical control to allow it to do its best job. For example, one machine-tool company regrinds all new standard milling cutters to a specific size before they

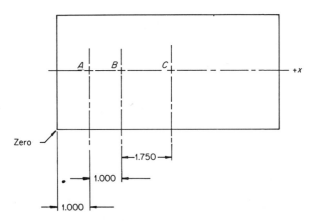

Fig. 3-6 Incremental or conventional dimensioning. Note that dimensions are not referenced to zero, except for point *A*.

get to a tool crib. N/C is not just a new automatic machine in a plant. It is a revolution and should be considered so. Traditional thinking must be changed or discarded entirely.

INCREMENTAL SYSTEMS

Incremental systems position the work or cutter from the point immediately preceding. In other words, calculations are made from where the tool or table is, to where it is going. A simple explanation is shown in Figs. 3-6 and 3-7, where only the x axis shall be used to demonstrate this system. Study the drawings.

Note that if a positioning error is made in the absolute system, then the next position is not affected by the error. In the incremental system the error is accumulative for all positions after the error. See Fig. 3-8.

The learners would do well to practice on a few similar exercises to become proficient in computing for both types of systems. To summarize, then, you have learned:

1. What is meant by positioning or point-to-point
2. To differentiate between absolute and incremental systems

Point	Increment
A	$x = 1$ inch
B	$x = 1$ inch
C	$x = 1\frac{3}{4}$ inches

Fig. 3-7 Notations for points A, B, and C in Fig. 3-6.

Fig. 3-8 A Wiedemann tape-controlled turret punch press with table travel of 1,000 inches per minute. An excellent example of a simple point-to-point system with absolute dimensioning. (*S.P.C. Machinery, Inc.*)

ZERO SHIFT SYSTEMS

Let us discuss the third consideration of the programmer: "Does the N/C machine tool have a fixed zero, a full-zero shift (or full-range offset), or a full floating zero?"

Cartesian coordinate system It is not the purpose of this section to point out advantages or disadvantages of any N/C machine, so only an explanation of the three basic types of zero shifts shall be made. To do this it may be well to understand something of cartesian coordinates and how they apply to understanding numerical control. A few simple charts will also help clear up any questions that may arise. René Descartes (1596-1650) is the man responsible for this very simple method of plotting points on intersections. Actually, this is one of his minor contributions, but a necessary one in N/C, to say the least. See Fig. 3-9.

NOTE:

1. There are four quadrants (numbered counterclockwise) around the zero point.
2. All units to the right of zero are plus x; all units above zero are plus y.
3. All units to the left of zero are minus x; all units below zero are minus y.

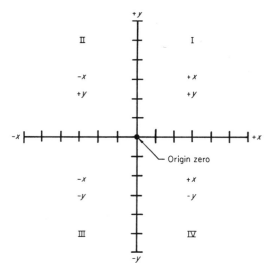

Fig. 3-9 Cartesian coordinate system showing standard designations for quadrants.

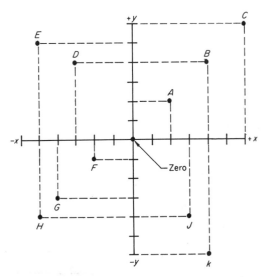

Fig. 3-10   Illustration showing points on intersections in the cartesian coordinate system.

Point	Quadrant	x	y	Point	Quadrant	x	y
A	I	+2	+2	F	III	−2	−1
B	I	+4	+4	G	III	−4	−3
C	I	+6	+6	H	III	−5	−4
D	II	−3	+4	J	...	...	...
E	II	−5	+5	K	...	...	...

Fig. 3-11   Solutions for points on intersections A through H in Fig. 3-10. Do J and K.

A few problems follow, and the values of some of the points have been calculated in the charts. Your instructor may have you complete these or similar problems on a separate sheet of graph paper. A little practice in establishing the values of points will be very helpful. Study Figs. 3-10 and 3-11.

If you now have confidence in your ability to plot points and are wondering how these simple problems relate to numerical control, the validity of these exercises will be demonstrated. You will actually be plotting values from a drawing of a part to be machined. Use the *absolute system* and note that the upper left corner has been chosen as the datum point. Programmers once felt that it was easier to program everything plus, plus; but experience has shown that it is just as easy to program in any quadrant if the N/C machine has the capability. Calculate values for x and y. See Fig. 3-12.

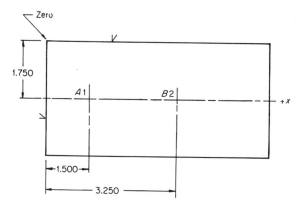

Fig. 3-12 Drawing of workpiece. Absolute dimensioning. See Fig. 3-13 for coordinate values.

Procedure

1. Decide which quadrant(s) is/are to be used.
2. If in doubt, go back and review quadrants.
3. Number or letter each position to be machined.
4. Now that you have done some review, plot the positions.
5. If you had no trouble with this problem, read on. If you had difficulty, practice with some similar problems.

Do *not* attempt to go on in this text until you become proficient in establishing these values from simple drawings. See Fig. 3-13 for solutions.

While the solutions may seem ridiculously easy to many people, there are those who, for one reason or another, may have difficulty. This text is written to help *all* those interested students who see the value of learning about numerical control.

If you have successfully mastered the absolute system, it is time to try an exercise using the incremental system. Use the same part in conventional dimensioning. It should be evident by now that the draftsman who knows what types of N/C equipment are installed should dimension his drawings accordingly. Communications in a plant should be wide-open for interaction between departments. Only good organization can accomplish this. In a technical-school situation the

Quadrant	Point	x	y
IV	A1	+1.500	−1.750
IV	B2	+3.250	−1.750

Fig. 3-13 Coordinates for part shown in Fig. 3-12. Absolute dimensioning.

drafting departments, electrical and electronic departments, math department, and machine shops should work together closely to develop a curriculum that is adequate, and that has numerical control as the pivot, without sacrificing any fundamentals of any of these areas.

Such a curriculum might include, in addition to machine shop, blueprint reading, shop mathematics, and courses in basic electricity, electronics, maintenance, and, possibly, Boolean algebra.

Procedure

1. Decide which quadrant is to be used. See Fig. 3-14.
2. Number or letter each position to be machined.
3. Plot the points for x and y (Fig. 3-15).

Now that a few positions have been calculated using the incremental system, it will be easier to understand the third consideration of the programmer. At this point can be reviewed what has been accomplished so far.

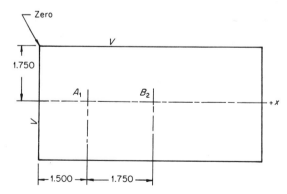

Fig. 3-14 Drawing of workpiece with conventional dimensioning. See Fig. 3-15 for coordinates.

1. You have an understanding of what a point-to-point system means.
2. You know now the difference between an absolute system and an incremental system.
3. Cartesian coordinates have been plotted from a reference point or zero, and you can establish which quadrant(s) you are working in. With this knowledge about cartesian coordinates you have actually plotted some points, assigning the correct values in the two axes, x and y.

It is important at this time that the student of N/C realize that, in order to calculate the points, a datum or zero reference point has been introduced into the problems. From this datum you have been able to calculate the distance necessary to travel in order to be positioned over a specific point. To oversimplify, you must know *where you are* before you can tell how *far* to go to reach any *destination,* and in what *directions* you must travel. Many N/C machine tools have a built in zero from which all calculations are made.

Fixed zero systems The zero point may be in a permanent position on a drilling or milling machine table, or it may be possible to shift it around manually with dials or electronic controls to any desired point. The important consideration here is that the programmer must know which type of reference he is concerned with at the time. These systems will be explained separately, hopefully in a clear and understandable way.

The first type or class of N/C machine tool may have what is known as a fixed zero with a small amount of shift possible in each axis. A machine tool of this type has a specific zero location on the table or way, as in the case of a lathe. Some adjustment of this point may be possible in either axis when necessary. All dimensions are then calculated from this zero to the locations of points on the workpiece

Quadrant	Point	x	y
IV	$A1$	$+1.500$	-1.750
IV	$B2$	$+1.750$	0.000

Fig. 3-15   Coordinates for incremental dimensioning in Fig. 3-14.

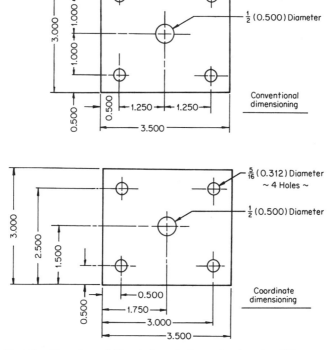

Fig. 3-16   Drawings of identical workpieces showing the difference between incremental and coordinate dimensioning.

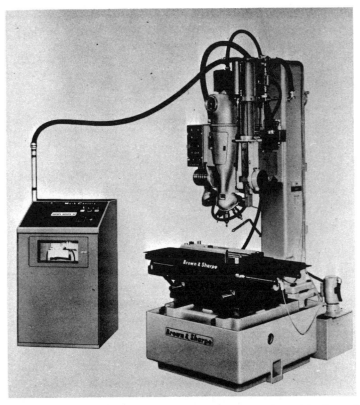

Fig. 3-17 Brown & Sharpe Turr-E-Tape N/C machine with General Electric Mark Century Control. (*Brown & Sharpe Mfg. Co.*)

wherever it is situated on the table. In practice, the dimensions on a drawing are recalculated to include the position of the machine zero. The work is then mounted on the table in the previously calculated position within the zero-shift capability of the N/C machine tool. All programmed dimensions are then in the plus, plus first quadrant with no provision for any minus readings. The following sketch (Fig. 3-18) will demonstrate this class of fixed zero.

It should be noted here that provisions for accurate sub-plates or grid plates have been made for easy location of the work in relation to machine zero.

Full zero shift or full-range offset The machine zero in this type can be adjusted to *any* point on the machine. Programmed dimensions must still be in the plus, plus first quadrant, however, with no minus reading capability. The workpiece can be placed anywhere, but all dimensions must be plus, plus (Fig. 3-19).

Full floating zero This type of zero shift does permit plus and minus programming from any selected zero point. See Fig. 3-20.

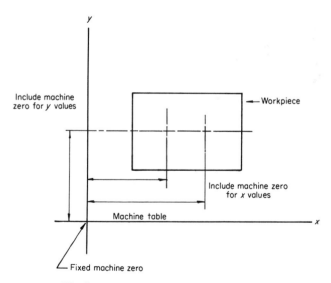

Fig. 3-18 Fixed-zero system. Illustration demonstrates first-quadrant programming (plus x, plus y).

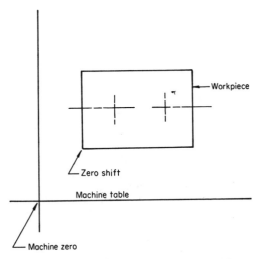

Fig. 3-19 Full-range zero offset. First-quadrant programming.

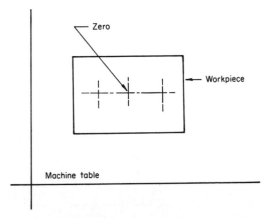

Fig. 3-20 Full floating zero. Demonstrates origin, and plus and minus programming.

SUMMARY

In this chapter, then, some of the things the process planner or programmer has to consider have been described. A list of these requirements should be a guide for beginners in the field. As a summary of this section, review what a programmer does and what he has to know about numerical control.

A programmer

1. Studies the blue-print
2. Chooses the N/C machine tool to be used
3. Identifies the type of material to be machined
4. Establishes the procedure for machining
5. Checks the tooling needed
6. Calculates the speeds, feeds, and other necessary considerations

The programmer must know:

1. Whether the system is point-to-point or continuous path
2. Whether the system is absolute or incremental
3. Which of three basic types of zero shift is used on the machine tool
4. The format for programming

This last consideration of the programmer shall be the subject of the next chapter.

REVIEW QUESTIONS

3-1. Describe a point-to-point system.

3-2. When the cutter is retracted and travels to a new position for drilling, is the path controllable by the operator or programmer in a positioning system?

3-3. Why do you think some industries destroy a duplicate tape after a job has been completed?

3-4. Can point-to-point systems be controlled on more than three axes?

3-5. Why should an operator of N/C machine tools be provided with complete instructions for running a job? Give at least three reasons.

3-6. Should blueprints of workpieces be suitable for the specific N/C system to be used in manufacturing the part? Why?

3-7. Of the two N/C systems, absolute and incremental, which one would you prefer to have in your shop? Why?

3-8. Which N/C zero-shift system would you prefer in your shop? Why?

3-9. On an N/C machine with a fixed zero, can a programmer write a manuscript using more than one quadrant? Illustrate.

3-10. What is an important disadvantage of an incremental system?

3-11. What does a programmer for N/C do?

3-12. What should a programmer know about the N/C systems in his organization?

4
Tape Format Classifications and Specifications

A technical definition of format classification states that the *format* is a means, usually in an abbreviated notation, by which the motions, dimensional data, type of control system, number of digits, auxiliary functions, etc., for a particular system can be denoted. This is an excellent description of format, although a little involved.

To clarify this definition, format means the language that a particular numerical control system understands and can act upon. If the control system can accept one kind of format, it cannot understand any other type; it will simply ignore any commands that it does not comprehend.

In order to learn about formats, some knowledge of punched tape and codes is necessary.

TAPE STANDARDS

The Numerical Control Panel of the Aerospace Industries Association (A.I.A.) and the Electronic Industries Association (E.I.A.) have done their homework well. The decade before 1959 had seen a proliferation of configurations for tape and punched cards. Machine tool manufacturers and their customers became justly concerned when they realized how confusing the field was becoming.

Differences in the physical dimensions of tapes and punched cards, along with varied coding practices, threatened to ruin what all knew to be a valuable concept. Even within one plant the varieties of input media would require a prohibitive number of people to be trained for the different systems. The differences in formats and coding alone were enough to cause everyone to take a breather and decide on a campaign to bring order out of the threatened chaos.

Standard RS-227 was issued by the Electronic Industries Association in October, 1959. This standard described a tape 1-inch wide with a maximum of eight rows of holes running longitudinally, and the diameters and spacing of these holes also standardized. This standard had the effect of immediately limiting the differences in tape preparation that were prevalent at that time. See Fig. 4-1.

National Aerospace Standard 943, released in January, 1960 by the A.I.A., set the hole patterns for letters and numbers. It is now a matter of record that 1-inch-wide punched tape is used in a majority of numerical control systems. (NOTE: The illustration of the E.I.A. standard coding for 1-inch-wide, 8-channel tape is in Fig. 4-2.) Specifications covered for punched tape include physical characteristics, mode and format requirements, material, and coding.

A complete coverage of these standards will be under-

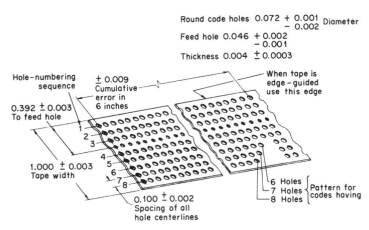

Fig. 4-1 Illustration showing physical characteristics of standard 1-inch punched tape. Standards NAS-943, NAS-955, and E.I.A. RS-227. (*Electronic Industries Association and Aerospace Industries Association of America*)

taken after some actual programming problems have been completed. A surface look at some usable tape information at this time, however, is in order to better understand format classifications.

READING THE TAPE

In Fig. 4-3 notice "binary counting 1-2-4-8." This information has been added to show how easily numbers can be read from punched tape. In this section of coded tape, see how this is done.

According to standard specifications, the total number of holes across the width of the tape must be odd. When the numeric code is even, a hole is punched in track 5. This is called a "parity check." For example, when the number 6 is punched, holes should appear in tracks 2 and 3. A hole

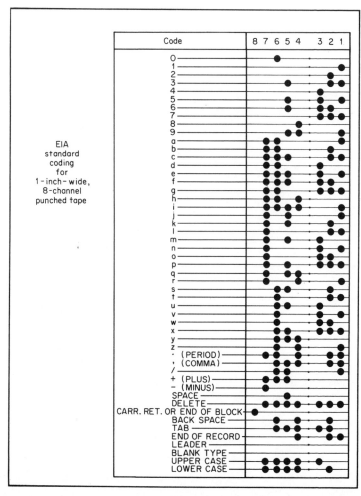

Fig. 4-2 Character codes for numerical machine-tool-control perforated tape. (*Electronic Industries Association*)

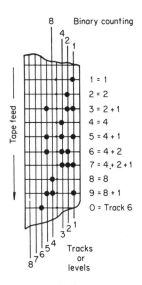

Fig. 4-3 Coding for numbers. Small dots are sprocket feed holes.

is also punched in track 5 to achieve an odd number of holes in that row.

Parity check will signal an error if a ragged or torn hole appears, or if there has been a coding error by the punch, such as an even number of holes for any character. It is really a safety device to help reduce the chances for error.

It might be well to note here also that some systems utilize an *even* parity check.

The United States of America Standards Institute has developed what is called the "United States of America Code for Information Interchange (USACII)." This coding system was developed to meet the increasing need for a recognized machine language code for interchange of information among information processing systems, communication systems, and other equipment. World-wide communications require more codes and symbols than are available in other standards.

This code does not affect the physical characteristics of tape, such as size, spacing of holes, etc. The N/C programmer need only be aware that this code is available and in use.

The numerals in USACII still have the same binary values as E.I.A. coding with these exceptions:

1. Holes are punched in tracks 5 and 6 to designate numerics (5) and lower case (6).
2. Where the number of holes is odd, a hole is punched in track 8 to establish *even parity*.
3. Track 7 is punched for upper case alphabetic characters, and tracks 6 and 7 for lower case alphabetic characters.

MANUAL PROGRAMMING

It appears that you are now ready to program for the first workpiece. In the introduction for this text it was pointed out that strict attention to detail was one of the requirements necessary to become a good programmer. Many operations performed on machines by skilled craftsmen are done instinctively, and with very little effort. If you have been in the machining field for any time at all, you know that just chamfering or burring a sharp edge on a workpiece is almost a conditioned reflex. The numerical control system however, must be directed in a clear and unmistakable manner to do this operation. The guide for programmers will be of great value now. A workpiece has been selected that appeared earlier in the text (See Fig. 4-4).

For this particular problem do not consider the type of material, or cutting speeds and feeds. It will be helpful at this stage to confine the problems to areas of numerical control with which the learner is now familiar. The purpose of this exercise, then, is to initiate some mental disciplines

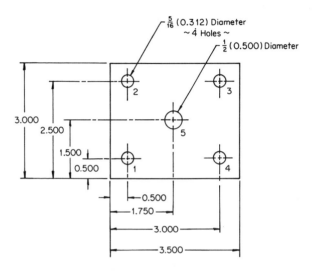

Fig. 4-4 Workpiece with coordinate dimensioning. This sample part will be programmed in four different formats and three types of zero shift.

and to start the potential programmer thinking in terms of coordinate values.

TAB SEQUENTIAL FORMAT

1. Study the drawing and number or letter the holes shown in Fig. 4-4. Consider positioning with the least amount of table movement. The holes have been numbered in a clockwise manner for this sample.
2. Use a numerically controlled drill press with two axes, x and y, under tape control, having these features:
    (a) Absolute dimensioning
    (b) Tab sequential format
    (c) Full-range zero shift (Fig. 4-5)
3. Use a center drill only for this program.

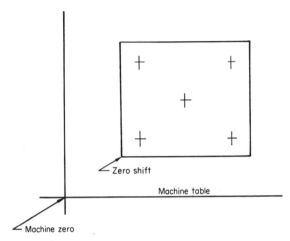

Fig. 4-5 Shows workpiece in Fig. 4-4 mounted on machine
table.   Operator shifts zero to lower left corner of part.

4. Do not consider work-holding clamps or devices for this
   problem.
5. Calculate the coordinate values for x and y (Fig. 4-6).

On the drawing shown in Fig. 4-5, the lower left corner
of the sample part has been selected as the zero reference.   All
dimensions, then, are in the first quadrant (plus x, plus y)
calculated from zero.

Notice in Fig. 4-6 that another set of coordinates has
been added.   After the coordinate values for hole 5 have been
calculated, the programmer would do well to plan on a work
loading and unloading area, free of encumbrances.   The
workpiece shall be programmed using the tab sequential
format.   This format utilizes a tab code as a spacer between
dimensions.   To simplify, when the tab key is depressed before the x dimension is typed, the coded holes on the tape
open a "gate" and the control accepts x information.   When

Hole #	x	y
• 1	0.500	0.500
2	0.500	2.500
3	3.000	2.500
4	3.000	0.500
5	1.750	1.500
Unload and load	5.000	5.000

Fig. 4-6 Coordinates for workpiece in Fig. 4-4, with full-range zero shift to lower left corner of part.

the tab code appears again in the tape, the x gate is closed and the y information is accepted. The following program exercise will demonstrate this feature.

In addition to knowing the systems involved in a particular machine-control unit there are details to be learned that should not be overlooked. In what order does the control read the numerical commands? For instance, how is .500 written if the control reads two whole numbers and three decimal places? For the first program, assume that the figures are read from right to left to establish the position of the decimal point. This means that trailing zeros must be included. All five places shall be used, however, to give practice and help you keep aware of this requirement. The decimal point does not have to be typed in the manuscript. The control knows where it should be.

The tab sequential format utilizes the tab character

to open and close gates for the reception of coordinate values for x and y. The person who prepares the tape does not depress x and y keys on the tape-punching equipment. The control knows that x information follows the first tab character, etc. An end-of-block character (a hole in track 8) must precede the first block of information to start the program, and end each succeeding block.

Undoubtedly there are many questions arising at this time. A few will be anticipated. Before this is done, however, a summary of what has been accomplished so far is in order. Coordinates have been plotted, tab sequential format has been introduced, and it has been pointed out that controls may search numerical values from right to left or left to right.

QUESTION How does the control know where on the work table the zero reference has been established so that the cutter can move from that point to the hole coordinates?

ANSWER Since one of the prerequisites for studying this text is knowledge of machine-shop practice, how the center of the spindle or cutter is positioned over the zero reference point on the workpiece shall not be gone into. This is a skill of the machinist, who may use edge finders, indicators, or specially designed fixtures for locating specific zero points on a machine table. However, once this physical operation is completed, the control receives this information from the operator in the form of push buttons on the control or similar devices that say to the control unit, "This is zero in the x axis. This is zero in the y axis." The control unit is now in possession of this information, and it is stored in its memory until a new zero reference is established or all information is cancelled. Essentially, this is the way a control knows where to go and in which direction from "home" or zero position. See Figs. 4-7 and 4-8.

Fig. 4-7 Control panel of a P & W Tape-O-Matic N/C drilling machine with milling capability. Photo shows switch for drill or mill modes (upper right). Buttons for setting zeros are also shown. (*Colt Industries Inc., Pratt & Whitney Inc., Machine Tool division*)

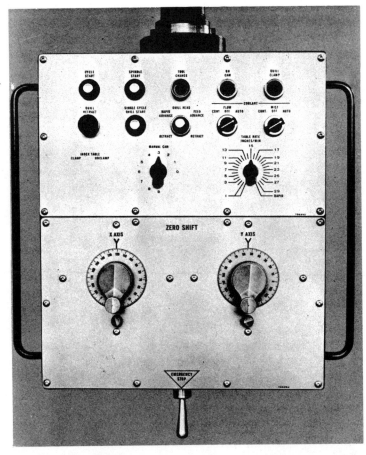

Fig. 4-8 Pendant station for horizontal and vertical Cintimatic machining centers. Shows levers for zero shift. (*Cincinnati Lathe & Tool Co.*)

SEQ #		x		y	
001	TAB	00500	TAB	00500	EOB
002	TAB	00500	TAB	02500	EOB
003	TAB	03000	TAB	02500	EOB
004	TAB	03000	TAB	00500	EOB
005	TAB	01750	TAB	01500	EOB
006	TAB	05000	TAB	05000	EOB

Fig. 4-9 Sample program manuscript for part in Fig. 4-4. Tab sequential format. Full-range zero shift. Long program.

QUESTION   In the sample program no plus signs have preceded the coordinate values. How does the control know whether the figures are plus or minus?   See Fig. 4-9.

ANSWER   Most MCU's in use today accept as plus any value having no sign. A minus sign must usually be punched in the tape, as indicated in some of the programming problems to follow.

QUESTION   Does the sequence number get punched into the tape?

ANSWER   Yes. In most controls there is a visual display of the block number being actively worked on. Three digits are allowed for sequence numbers and may be utilized in any manner desired for the personal information of the programmer, or for other reasons. Since these numbers have no other significance, many plants have devised codes to aid

Fig. 4-10 General Electric Mark Century Control. Note display of the sequence number in window at top of control. (*General Electric Company*)

them in many ways. For example, 200 might mean that cutter number 2 is in position at the moment.

QUESTION An end-of-block character (a hole in track 8) is mentioned, which must precede the first block in the program and end each block. What is the function of this code?

ANSWER The end-of-block code preceding the start of a program arms the parity-check circuits. This simply signals the control system that a program is about to begin. Parity check will signal an error, if one should appear on the tape in the form of an even number of holes in a row or a torn or ragged hole. This code at the end of a block of information simply defines the end of data for that block. On most tapepunching typewriters the carriage-return key, when depressed, will punch this hole in track 8.

Many such questions will be answered throughout the text. If you still have some at this point, it is entirely possible that they will be answered in subsequent passages.

Should some unfamiliar terms appear occasionally, it might be helpful to refer to the Glossary.

In the program manuscript just completed, no mention of setup instructions for the operator has been made, nor has any extraneous information been included, such as functions other than simply positioning the cutter over the hole locations. The purpose in doing this is to help the learner progress in logical steps from the simple to the more complex. These are only exercises at this point. The goal is to finally complete a usable program, one that shall have all the necessary information and be technically complete.

PREPARING THE TAPE

The physical preparation of the program tape is a simple process. A standard Friden Flexowriter* or Dura typewriter equipped with a punch unit are machines used for tape preparation. See Figs. 4-11 and 4-12. There are many types of machines on the market today for this kind of application. When a specific key is depressed the punch unit automatically cuts the correct codes in the tape, including the parity-check code as needed. Fig. 4-13 shows a sample punched tape.

The operators of these devices need only a minimum of practice in order to be able to copy program manuscripts. He or she needs no special knowledge, and, in fact, may even be a typist in an office of the plant, with no skills in programming.

* A trademark of Friden Inc.

Fig. 4-11 Numerical control tape-preparation center. Dura Mach 10-B automatic typewriters with tape output and input. Dura Bi-Directional converter system. (Logic boards are wired to accommodate various N/C machine code logic and code language requirements.) (*Dura Business Machines, division of Dura Corporation*)

A PROGRAMMING TECHNIQUE FOR TAB SEQUENTIAL FORMAT

Before leaving this tab sequential format, consider a programming technique that reduces the amount of typing and helps to shorten the program tape. Most numerically controlled systems utilizing the tab sequential format, have the capability of storing certain items of information. As long as there is no change in specific axis coordinates, there is no need to repeat the numerical information. It is only necessary to depress the tab key for that column.

Study the program manuscript (Fig. 4-14) and note how the information has been programmed. The first tape had 97 rows of information (Fig. 4-13). The second illustration has only 81 rows (Fig. 4-15).   While the savings in time and tape length appear to be negligible here, it can mean a great sav-

Fig. 4-12 Model 2301 Flexowriter automatic writing machine by Friden. (*Friden Inc., division of the Singer Company.*)

ings on complex jobs with many coordinates and miscellaneous functions.

FIXED-BLOCK FORMAT

There are really only two major differences between the tab sequential and fixed-block formats. They both require that numerical information follow a specific order, with these variations:

1. All words must be completely coded in fixed block, whereas in tab sequential words may be omitted as demonstrated previously.

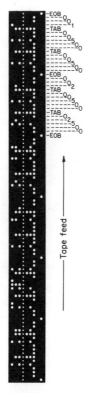

Fig. 4-13 Tape punched for manuscript in Fig. 4-9. Tab sequential format. Long program.

SEQ #		x		y	
001	TAB	00500	TAB	00500	EOB
002	TAB	. . . . .	TAB	02500	EOB
003	TAB	03000	TAB	. . . . .	EOB
004	TAB	. . . . .	TAB	00500	EOB
005	TAB	01750	TAB	01500	EOB
006	TAB	05000	TAB	05000	EOB

Fig. 4-14 Sample program manuscript for tab sequential format. Repeated words omitted from program. Tab key depressed. Short program.

2. Since the controls for fixed block are designed to accept words (all words, none omitted) in a specific order, there is no tab code used. The MCU must have all necessary information in order to function properly. This difference shall be demonstrated when programming the same workpiece as was used for the previous manuscript.

Procedure

1. Study the drawing (Fig. 4-4).
2. Number or letter the coordinate points to be considered.
3. Use a numerically controlled drilling or milling machine, two axes (x and y), under tape control having the features of absolute dimensioning, fixed-block format, and fixed zero.
4. Use a center drill only for this problem.
5. Do not consider work-holding devices for this programming exercise.

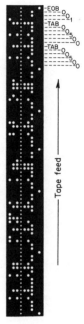

Fig. 4-15 Tape punched for Fig. 4-14. Short program. Tab sequential format.

6. Calculate the coordinate values for x and y (Fig. 4-18). It is now necessary to include the fixed zero in your solutions for the coordinates. The dimensions in Fig. 4-17 will demonstrate this condition for programming.

7. This control will read two whole numbers and three decimal places. See Figs. 4-19 and 4-20.

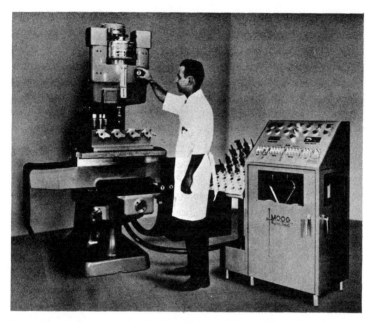

Fig. 4-16 A three-axis vertical milling machine. Fixed-block format programming for x, y, and z axes. (*Hydra-Point division, Moog Inc.*)

WORD-ADDRESS FORMAT

In this format words need not be presented in any specific order. Aside from this consideration there are some specific differences when compared to tab sequential or fixed-block formats. Each word in a block must be identified as to its meaning. For instance, $x02555$ is identified as referring to the x character. These identification characters must be punched into the tape. Many control systems are utilizing this format with some minor variations. Use the same workpiece for the next problem (Fig. 4-4).

For this programming exercise use a drilling or milling machine, two axes (x and y), under tape control having these

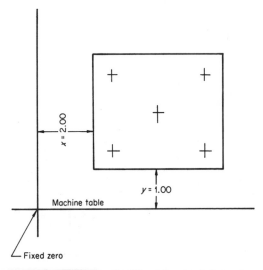

Fig. 4-17 Workpiece in Fig. 4-4 mounted on machine table having a fixed zero and showing part offset 2 inches in x axis and 1 inch in y axis from zero.

features:

1. Absolute dimensioning
2. Word-address format
3. Full floating zero (this will have given you practice in programming for the three types of zero-shift systems).

Procedure

1. Study the drawing (Fig. 4-4).
2. Number or letter the holes to be drilled.
3. Use a center drill only for this problem.
4. Do not consider work-holding devices at this time.

Hole #	x	y
1	2.500	1.500
2	2.500	3.500
3	5.000	3.500
4	5.000	1.500
5	3.750	2.500
Unload and load	7.000	6.000

Fig. 4-18 Coordinates for workpiece as mounted in Fig. 4-17.

SEQ #	x	y	
001	02500	01500	EOB
002	02500	03500	EOB
003	05000	03500	EOB
004	05000	01500	EOB
005	03750	02500	EOB·
006	07000	06000	EOB

Fig. 4-19 Program manuscript. Fixed-block format. Fixed zero. (For workpiece in Fig. 4-4.)

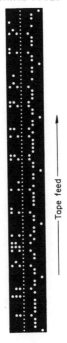

Fig. 4-20 Punched tape for program manuscript in Fig. 4-19.

Hole #	x	y
1	−1.250	−1.000
2	−1.250	1.000
3	1.250	1.000
4	1.250	−1.000
5	0.000	0.000
Unload and load	5.000	5.000

Fig. 4-21 Coordinates for workpiece in Fig. 4-4 with hole 5 as the zero reference.

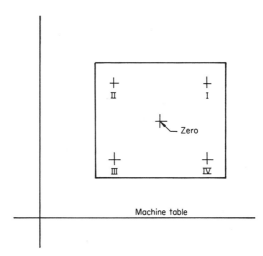

Fig. 4-22 Workpiece in Fig. 4-4 on machine table. Full floating zero. Center of hole 5 has been selected as the zero reference for the control. Plus and minus programming example. Drawing shows quadrants.

SEQ #			
N001	$x - 012500$	$y - 010000$	EOB
N002	$x - 012500$	$y010000$	EOB
N003	$x012500$	$y010000$	EOB
N004	$x012500$	$y - 010000$	EOB
N005	$x000000$	$y000000$	EOB
N006	$x050000$	$y050000$	EOB

Fig. 4-23 Program manuscript for word-address format. Full floating zero. All digits to be punched in tape. Long programs.

Tape feed

Fig. 4-24 Punched tape for program manuscript in Fig. 4-23. Word-address format. Long program. All words punched in tape even when repeated (119 rows).

SEQ #			
N001	$x - 0125$	$y - 01$	EOB
N002	. . . . .	$y01$	EOB
N003	$x0125$	. . .	EOB
N004	. . . . .	$y - 01$	EOB
N005	$x0$	$y0$	EOB
N006	$x05$	$y05$	EOB

Fig. 4-25 Program manuscript. Word-address format. Trailing zero suppression. Repeated words not punched.

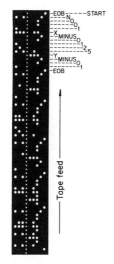

Fig. 4-26 Punched tape for program manuscript in Fig. 4-25. Trailing zeros suppressed. Repeated words not punched (63 rows of information).

SEQ #					
N001	TAB	x02500	TAB	y01500	EOB
N002	TAB	......	TAB	y03500	EOB
N003	TAB	x05000	EOB	......	....
N004	TAB	......	TAB	y01500	EOB
N005	TAB	x03750	TAB	y02500	EOB
N006	TAB	x07000	TAB	y06000	EOB

Fig. 4-27 Program manuscript for Fig. 4-4. Variable block format. Fixed zero.

5. Calculate the coordinate values for x and y. The center of hole 5 has been chosen for the zero reference point. Program in all four quadrants. There will also be some calculations to make, since all holes are to be referenced from the center of hole 5 (Fig. 4-21).

6. Visualize the quadrants, the quadrant numbers having been added to the drawing of the workpiece mounted on the machine table (Fig. 4-22).

7. Assume that this control system reads two whole numbers and four decimal places reading from left to right. Many control systems that read the numeric values from left to right to establish the decimal point are designed so that trailing zeros, which are not significant, may be left out of the tape. In addition, repeated words need not be punched. Study Figs. 4-23 to 4-26.

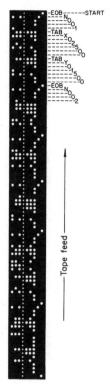

Fig. 4-28 Punched tape for program manuscript in Fig. 4-27. Variable block format.

VARIABLE BLOCK FORMAT

A combination of the word-address format and tab sequential has resulted in what is called a "standard compatible." There is still much research going on in this area of formats. Differences that now exist preclude the possibility of using

Fig. 4-29 Carlton horizontal N/C drilling machine. Four-inch spindle. Has boring and milling capabilities. (*S.P.C. Machinery, Inc.*)

a program tape from one MCU on another control unit unless the control is identical.

Variations in formats are really no problem for an experienced programmer. The differences are shown in this chapter so that the beginner will understand the four basic types, and, thus, easily adapt to any future changes.

For this program use a milling or drilling machine, two axes (*x* and *y*), under tape control having these features:

1. Absolute dimensioning

2. Variable block format (standard compatible)
3. Fixed machine zero; coordinates to be the same as given in Fig. 4-18.

Follow the procedure used in previous problems. This control reads two whole numbers and three decimal places from right to left. Compare your coordinate values with those in Fig. 4-18. Study the program manuscript (Fig. 4-27). Read the tape (Fig. 4-28).

If you have followed these exercises carefully and are now confident that you understand what has been presented, then you are well on the road to further knowledge in this field of N/C.

Many variations of these formats and control systems are in existence, but the basic principles will be the same. When going on to contouring controls, incremental dimensioning will be used to give practice in this area.

SUMMARY

The accomplishments of this chapter are:

1. Coordinates have been plotted.
2. Program manuscripts have been written for fixed machine zero, full-range zero shift, and full floating zero.
3. Absolute dimensioning has been used with all of the problems.
4. Four formats—tab sequential, fixed block, word address, variable block—have been used.
5. Numeric values have been written for controls that search right to left and left to right.
6. Tape specifications and preparation have been discussed.

It should be of interest to note that, although the same workpiece has been used for four programming problems, the three different zero shifts used caused the solutions for the coordinates to be different for each system.

REVIEW QUESTIONS

4-1. Why did the Aerospace Industries Association and the Electronic Industries Association decide to standardize tapes and tape coding?

4-2. Specifications for punched tape describe what four characteristics?

4-3. What is meant by odd parity?

4-4. What is meant by even parity?

4-5. Does USACII coding affect the physical characteristics of standard tape?

4-6. In E.I.A. coding, which track number is used for the parity check hole?

4-7. In USACII coding, which track number is used for the parity check hole?

4-8. In the tab sequential format does the tab code have a specific function? Explain.

4-9. Does a decimal point get typed in the programmer's manuscript? Why?

4-10. Does the person who punches the tape for N/C have to punch the parity hole specifically?

4-11. What are some of the reasons for the development of USACII coding?

4-12. Why is it necessary to depress the carriage-return key first when punching tapes for many N/C systems?

4-13. Is it necessary to be a programmer in order to punch N/C tapes? Why?

4-14. What is one advantage of not having to repeat information on tape when there is no change in coordinates from a previous block?

4-15. In the fixed-block format does the tab key have to be depressed for punching tape? Explain.

4-16. How does the word-address format differ from the tab sequential and fixed-block formats?

4-17. If the programming formats are different, can a tape in one format be used on a control that accepts a different format?

4-18. Which formats resemble the variable block (standard compatible) format?

4-19. Of the different formats you have studied, which do you prefer? Why?

5

Functions Other Than Machine Table or Spindle Movements

Earlier in the text it was mentioned that milling operations were a significant part of numerical control applications. To make this operation readily understood, a milling mode will be included in the next programming problem. Before doing this, however, some of the standard functions included in many machine control units will be discussed. Up to this point a cutter (center drill) has been positioned over the centers of holes to be drilled. Visualize some of the operations that have to be performed in order to drill these holes, such as changing tools, turning the spindle on, advancing the tool into the workpiece, selecting feed rates, turning coolant on or off, etc.

Many control systems include these auxiliary and miscellaneous functions as part of the control unit package to be punched into the program tape. The more functions on tape, the fewer operations performed by the operator. It has been stated that some N/C equipment is so designed that the operator simply loads and unloads workpieces. While the machine is producing a new part, he may be loading or unloading another machine, or at least performing some operation such as burring or checking finished parts. The tape standards and specifications that were discussed previously also include specifications for functions other than axis dimensions. An explanation of some of these functions is in order at this time.

SEQUENCE NUMBER, N, PLUS THREE DIGITS

This number has been explained previously. It has no significance other than to identify, usually in a visual display, the block number in active command on a specific program tape. It is only common sense then, that it be the first word in a block.

PREPARATORY FUNCTIONS, G, PLUS TWO DIGITS

This function merely instructs the machine to get ready to perform a specific type of operation. As an example, if the programmer is going to call for an arc to be cut in a clockwise direction, he may punch a G02 in the tape. The preparatory function precedes a dimension movement. This also makes sense because the machine has to know what type of operation it is going to perform. Preparatory functions have two digits assigned to describe the operation. They may range from G00 through G99.

FEED RATES, F, PLUS THREE DIGITS

Machine control unit manufacturers have numerous methods for designating feed rates. It should be noted here that the standards set by the E.I.A. and A.I.A. are guidelines and some variations will appear. The programmer must know the specific control unit with which he is working. Feed rates are specified in inches per minute. Magic 3 codes and feed rate numbers in addition to direct feed rates will be covered in a later chapter.

SPEED CODES, S(rpm), PLUS THREE DIGITS

Speeds of spindles or cutters are coded in many ways by control manufacturers. Here again, knowledge of the specific MCU is essential. Some controls may utilize proprietary codes in this area.

TOOL SELECTION, T (AUTOMATIC CHANGERS), PLUS AS MANY AS FIVE DIGITS

The tool word, as in the other functions, may vary according to the control design. However, the T plus a code number brings a specific tool into position.

MISCELLANEOUS FUNCTIONS, M, PLUS TWO DIGITS

This function pertains to information such as spindle start, spindle direction, coolant on or off, spindle stop, and other functions not specifically part of the cutting portion of the program.

AXIS DESIGNATIONS

It is not the purpose of this text to introduce more than two axes, x and y. Angular dimensions around x and y, plus other axes, are subjects for possibly another, more advanced

Fig. 5-1 Brown & Sharpe Hydrotape N/C machine. Tool index into position called out on tape. (*Brown & Sharpe Mfg. Co.*)

text. If the learner can just become proficient and confident in his ability to program simple machines at first, then his logical progression to more complex programming is assured.

Finally, there are a few letter address characters specified for other functions, but they will not be covered at this time. One or two of these will appear when programs for a contouring or continuous path numerically controlled machine tool will be studied.

GUIDELINES FOR PREPARATORY FUNCTIONS

The following codes have been assigned to the G functions. It should be stressed here, however, that the programmer may be concerned with control units that do not adhere to these suggested codes. Freedom of design is, of course, the manu-

Fig. 5-2 Cincinnati Cintimatic 8 spindle turret drill. Closeup of turret shows tool station numbers called out on tape for indexing into position. (*Cincinnati Lathe & Tool Co.*)

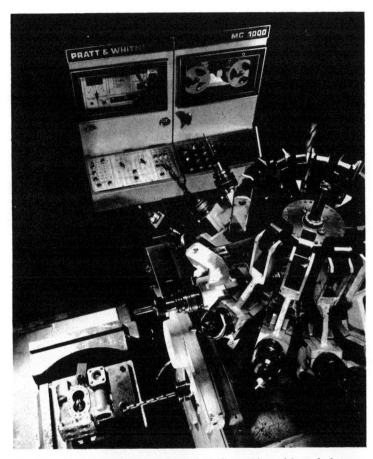

Fig. 5-3 Pratt & Whitney horizontal N/C machine with tool changer. (*Colt Industries, Inc., Pratt & Whitney, Inc., Machine Tool division*)

facturer's prerogative, but most controls will follow pretty closely the standard specifications.

Of a possible 100 combinations (G00 through G99), only 48 codes have been assigned as of now.   Of the 48 assigned codes, 14 of these are for control use only, or for cutter compensation use.

To keep this section of the text uncluttered, only those codes that have been assigned by the standards associations, with the exception of codes for control use only and cutter compensation codes, will be explained.   See Fig. 5-4.

GUIDELINES FOR MISCELLANEOUS FUNCTIONS, M, PLUS TWO DIGITS

It was explained earlier that miscellaneous functions are those that are not specifically a part of the *cutting* program.   These codes have been assigned a letter address, M, and two digits. Of a possible 100 codes only 32 have been assigned at this writing.   Of these 32 M codes, 4 have been reserved for control use only.

Consider only the remaining 28 miscellaneous functions.

Selection of tape-controlled feed rates, speeds, tool indexing, etc., will be included in the contouring controls section.

A milling operation, utilizing a few of the preparatory functions with some miscellaneous functions added, should be of great help to the novice programmer at this point. Everything learned up to now plus these added functions will keep things in proper perspective (Fig. 5-5).

MILLING OPERATIONS, POINT-TO-POINT

Some controls simply utilize a switch after the milling cutter is clamped in cut position.   The operator initiates the cutting action by changing the aforementioned switch from drill to mill and selecting manually the milling feed rate in inches per minute.   Other control systems are designed so that these

Code	Function
G00	Used with combination positioning/continuous path systems. Signifies a positioning operation.
G01 G10 G11	Describes a linear interpolation block in systems that need this description.
G02 G20 G21 G03 G30 G31	Assigned for circular interpolation.
G04	Dwell code.
G05	Hold. Machine motion halted.
G08	Acceleration code.
G09	Deceleration code.
G13 G14 G15 G16	Prepares system to operate on a specific axis.
G17 G18 G19	Selects plane for circular interpolation or cutter compensation that operate in two dimensions simultaneously.
G33 G34 G35	Mode for threading, constant lead. Mode for threading, increasing lead. Mode for threading, decreasing lead.
G80 through G89	Assigned for fixed cycles.

Fig. 5-4 Partial list of codes for preparatory functions. E.I.A. Standard RS-273, Interchangeable Perforated Tape Variable Block Format for Positioning and Straight Cut Numerically Controlled Machine Tools. (*Electronic Industries Association*)

Code	Function
M00	Machine stop.
M01	Optional stop. Operator signals previous to block by push button.
M02	End of program. May rewind tape.
M30	Will rewind at end of tape.
M03	Spindle on, clockwise.
M04	Spindle on, counterclockwise.
M05	Stop spindle.
M06	Tool change (not selection) manually or automatically.
M07	Coolant on (possibly flood).
M08	Coolant on (possibly mist).
M09	Coolant off.
M10 M11	Clamp code ⎱ Spindles, fixtures, slides, etc. Unclamp code ⎰
M13	Clockwise-spindle on–coolant on.
M14	Counterclockwise-spindle on–coolant on.
M15 M16	Rapid traverse or feed motion ⎱ Plus M15 ⎰ Minus M16
M31	Interlock bypass—circumvent normal interlock.
M32 through M35	Maintain constant surface cutting speed.
M40 through M45	Gear change codes.

Fig. 5-5 Partial list of codes for miscellaneous functions. E.I.A. Standard RS-273. (*Electronic Industries Association*)

functions are completely tape controlled. Both are excellent systems, and only the in-plant needs of the machine buyer will dictate which is preferable for his organization.

For this milling program exercise, use a vertical milling machine or drilling machine with milling capability, having two axes, x and y, under tape control with these features:

   I. Absolute dimensioning.
  II. Word-address format.
      A. Repeated words may be omitted.
      B. Control reads two whole numbers and four decimal places from left to right to establish the decimal point. Trailing zeros may be dropped. Leading zeros must be programmed.
 III. Full floating zero.
  IV. Preparatory functions on tape.
      A. When the table of assigned preparatory function codes is studied, note that G80 through G89 have been designated as recommended for fixed cycles (sometimes called canned cycles). These codes do not include a milling cycle. A fixed, or canned, cycle is a series of preset (designed into the control) operations that direct machine axis movement and/or spindle operation to complete boring, drilling, tapping, or, in this case, milling operations. For our hypothetical MCU, then, some recommended positioning codes from E.I.A. Standard RS-274-A will be used to initiate the following series of movements (fixed cycle).
      1. G75—Rapid traverse the center of the cutter over the desired start-cut position.
      2. Rapid traverse down to a predetermined distance from the actual depth of cut.
      3. Go to depth as shown on B/P in the manually selected feed rate.

4. Stop motion. At this point the operator will clamp the quill. (Some controls have an automatic quill clamp here.)

B. G65—Preparatory function to initiate the actual milling operation. Note that this code is chosen so as to tell the machine to start the cutting operation when the "cycle start" button is depressed. These codes have been chosen for a hypothetical control and are not intended to represent any specific N/C unit.

V. Miscellaneous functions on tape.

A. After the milling operation has been completed, it might be useful to design the MCU to accept a miscellaneous code such as a program stop. This code will stop the spindle and coolant flow. When the table of assigned codes for miscellaneous functions is studied, note that M00 is applicable. Utilizing this code will allow the operator to unclamp the quill, retract manually, possibly change to another tool if needed, and continue with the program.

B. A coolant-on code in the control might be included (M08). For the purposes of this exercise, other functions shall be manually set. These include spindle start, feed rates, speed (rpm), and retract quill. Many variations of control units are available. The programmer must know his machines and controls. These simple illustrations should help the beginner to be aware of the things he must know.

VI. Format detail—Program in this order:

N	G	x	y	M
Three digits	Two digits	00.0000	00.0000	Two digits

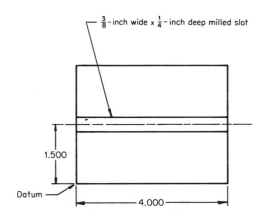

Fig. 5-6 Sample part for milling operation.

Procedure

1. Study the drawing (Fig. 5-6).
2. In Figure 5-7 the milling cutter has been added at the start-cut position A, and end-cut position B.

    Scale drawings of clamps, fixtures, workpieces, grid plates, vises, and cutting tools etc., in position, are all used frequently when programming for numerical control. Standard tools, grids, etc., of plastic are very helpful. Do not ever hesitate to plan mockups or visual aids if you feel their efficient use will expedite the programming. A hydraulic tool turret indexing at 300 psi can cause much damage if there is insufficient clearance.
3. Calculate the coordinates for:
    a. Cutter position A—the center-line of the cutter. Allow $\frac{1}{8}$ inch for approach.
    b. Cutter position B (Fig. 5-8).
4. Do not consider work-holding devices at this time.
5. Prepare a manuscript (Fig. 5-9).
6. Prepare a tape.

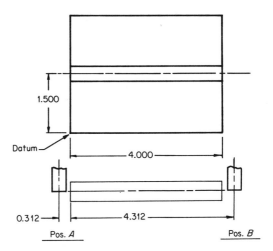

Fig. 5-7 Milling cutter shown at start-cut, end-cut positions. To be programmed in word-address format. Full floating zero.

PROGRAMMING A COMPLETE JOB: POINT-TO-POINT

One of the attributes of an efficient programmer is the ability to look at a drawing of a workpiece and visualize it in a special way. He must see the part as a whole entity, and, at the same time, be able to process the operations mentally

	x	y
A	-0.3125	1.500
B	4.3125	1.500

Fig. 5-8 Coordinates for milling cutter as shown in Fig. 5-7.

N	G	x	y	M	Operator
N007	G75	$x-003125$	$y015$	. . . .	Clamp quill
N008	G65	$x043125$	. . .	M08	Initiate milling operations
N009	. . .	. . . . . .	. . .	M00	Unclamp quill— retract

Fig. 5-9 Program manuscript for workpiece in Fig. 5-6. Word-address format. Full floating zero.
NOTE: Milling codes have not been standardized. Some N/C machines use G78, G79. There are actually many codes used for this operation. The suggested standards are guidelines, but manufacturers may deviate from the suggested codes.

in their proper order. By this is meant the elimination of all unnecessary movements of the table and carriage. Decisions as to which cutting operations should be done first and which are efficient speeds, feeds, etc., all flash through the programmer's mind as he studies the drawing. Studying a drawing does not mean just looking at it, however. Only after studying the drawing does the programmer begin to physically plan the program.

To program a complete workpiece will require the reader's complete and undivided attention. Plan your time so that you will have at least 1 to 2 hours of complete application to the following programming problem. At this point you will be using all of the information learned to date with a minimum of guidance. The solutions will, of course, be included with an added type of document called a tool sheet. It is possible that you may have to keep referring back to information in this text.

The aluminum workpiece selected for this programming exercise includes five drilled holes plus a milled slot. To give the beginning programmer maximum practice, centerdrilling, drilling, and milling operations will be done. Many shops are using stub- or spiral-point drills, which do not require spotting. However, the more operations programmed at the beginning, the easier it will become to utilize shortcuts. The machining of the workpiece is not the final objective at this time. It is merely a means to an end. All of the ingenuity you can muster under actual plant conditions will enhance your career in numerical control. It might be interesting to devise your own program from Fig. 5-10 after you have followed through in the text.

An expert programmer must have infinite patience and a solid background in machining operations. Knowledge of all of the available codes, formats, control systems, and computer assists should be his goal. Constant and thorough research, and reading of trade manuals will keep an efficient programmer up to date.

Procedure

  I. Study the drawing (Fig. 5-10).
 II. Number or letter the holes and slot in the selected machining sequence.
III. Use a numerically controlled vertical milling machine or drilling machine with milling capabilities having the following features.
   A. Two axes, x and y, under tape control.
   B. Absolute system.
   C. Tab sequential format.
   D. Full floating zero, plus and minus programming.
   E. Controls read two whole numbers and four decimal places.
     1. Repeated words need not be programmed.
     2. Leading zeros must be programmed.
     3. Trailing zeros may be suppressed.

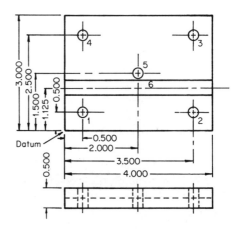

NOTE : Holes $-\frac{1}{4}$ (0.250) Diameter No. 1, 2, 3, 4. 5
Milled slot $-\frac{3}{8}$ - inch wide $\times \frac{1}{4}$ - inch deep No. 6

Fig. 5-10 Sample workpiece to be programmed. Tab sequential format. Full floating zero.

F. Preparatory functions on tape, G

   1. 81—Drill cycle: rapid traverse to position in x and y, down to gage height, feed to depth, quill retracts.

   2. 77—Mill cycle: rapid traverse in x and y to position, down to gage height, feed down to position, motion stops, operator clamps quill.

   3. 66—Initiate milling operation to end of cut.

   4. 80—Cancel cycle. Rapid traverse to position in x and y.

G. Miscellaneous functions on tape, M

   1. 06—Tool change.

   2. 36—Tool change and tape rewind.

   3. 51, 52, 53—Manually set depth cams for rapid traverse down, and feed to depth.

Fig. 5-11 Closeup photo showing drum cams for manually setting depth for rapid traverse and feed rate. (*Cincinnati Lathe & Tool Co.*)

    4. 08—Coolant on.
    5. 09—Coolant off.
IV. Prepare tool sheet (Fig. 5-13).
 V. Prepare manuscript. See Figs. 5-14 and 5-15.
VI. Prepare tapes: one master and one duplicate.

FORMAT DETAIL

Control systems are described in a special way. Format detail information describes the words used and also the order in which they must appear on the tape. To describe the hypothetical format detail to be used in this exercise, the

Fig. 5-12 This illustration shows how more than one set of drum cams may be used when many tools are needed on one job.   The whole assembly may be quickly taken off or put on the machine.   (*Cincinnati Lathe & Tool Co.*)

Tool sheet		XXX Machine Co.		
Part no. XXX		Fixture—Vise—-Tool no. XXXX		
Part name XXX				
		Machine—Pt-to-Pt no. XXXX		

Depth cam number	Procedure	Tool description	Spindle	
			Speed	Feed
1	Center drill (five holes)	E2	2,000	0.003
2	Drill five holes through	$\frac{1}{4}$-inch drill	2,500	0.003
3	Mill slot	$\frac{3}{8}$-inch end mill	2,000	4 ipm

Fig. 5-13 Tool sheet for planning program of part in Fig. 5-10.

information would be written in this manner:

N3.G2.x2.4.y2.4.M2.M2*

where . denotes the tab character, and * denotes the end of block character.

Explanation

N3 represents the sequence number (three digits).

G2 represents the preparatory function (two digits).

x2.4 indicates there are two digits to the left of the decimal point and four digits to the right for the x dimension.

y2.4 is the same as above, except this is for the y dimension.

M2 denotes the miscellaneous function (two digits).

NOTE: Another M circuit has been added if it is desired to use more than one miscellaneous function in a block of information. Some controls allow many M functions to be programmed in a single block as long as one does not contradict another, e.g., spindle off and spindle on.

Hole number	x	y
1	0.500	0.500
2	3.500	0.500
3	3.500	2.500
4	0.500	2.500
5	2.000	1.500
Tool change	7.000	6.000
Coordinates for milling cutter		
A	-0.500 ($\frac{5}{16}$ approach)	1.125
B	4.500	1.125

Fig. 5-14 Coordinates for holes and milling cutter in Fig. 5-10. Full floating zero.

Explanation

The following is an explanation of the tape program in Fig. 5-15.

000—Set up information. Operator zeros machine and manually sets rpm, sets feed.
001—Drill cycle. Coolant on. Stop 1.
002—Rapid traverse to new x dimension. Drill cycle.
003—Rapid traverse to new y dimension. Drill cycle.
004—Rapid traverse to new x dimension. Drill cycle.
005—Rapid traverse in x and y to hole position 5. Drill cycle.

N	T	G	T		x	T		y	T E	M	T E	M	T E	Operator
000	T		T		0	T		0	E					Setup
														information
														Zero − B/P
														rpm
														Feed
001	T	81	T		005	T		005	T	51	T	08	E	Set stop #1
002	T		T		035	T			T		T		E	
003	T		T			T		025	T		T		E	
004	T		T		005	T			T		T		E	
005	T		T		02	T		015	T		T		E	
006	T	80	T		07	T		06	T	06	T	09	E	Tool change
														rpm
007	T	81	T		005	T		005	T	52	T	08	E	Set stop #2
008	T		T		035	T			T		T		E	
009	T		T			T		025	T		T		E	
010	T		T		005	T			T		T		E	
011	T		T		02	T		015	T		T		E	
012	T	80	T		07	T		06	T	06	T	09	E	Tool change
														rpm
														Feed
013	T	77	T	−	05	T		01125	T	53	E			Clamp quill
														Stop #3
014	T	66	T		045	T			T		T	08	E	
015	T	80	T		07	T		06	T	36	T	09	E	Tool change
														Unclamp quill
														Unload
														Load

Fig. 5-15 Program manuscript for complete workpiece in Fig. 5-10.   Tab sequential format, Full floating zero, trailing zero suppression, repeated words tabbed.

006—Cancel cycle. Rapid traverse to tool change coordinates. Spindle and coolant stop. Change to ¼-inch drill. Manually set new speed and feed.

007—Traverse in x and y and repeat drill cycle depth stop 2.

008—Same as 002.

009—Same as 003.

010—Same as 004.

011—Same as 005.

012—Cancel cycle. Rapid traverse to tool change position. Insert ⅜-inch mill cutter. Manually set rpm and feed rate.

013—Mill cycle. Operator clamp quill at depth.

014—Milling operation initiated. Coolant on.

015—Cancel cycle. Rapid to unload position. Unclamp quill. Retract. Load next workpiece. Change tool. Tape rewinds.

NOTE: Some controls provide for a tape-rewind stop code over first operation coordinates instead of over zero as in this program.

PROGRAM: FIXED-BLOCK FORMAT

Many MCU's in use at this time are designed to accept the fixed-block format. This is a relatively inexpensive system and has found wide acceptance. All necessary information must be programmed, even though repeated in subsequent blocks.

Program the same workpiece (Fig. 5-10) and choose a vertical milling machine with two axes, x and y, under tape control having these features:

1. Absolute system.
2. Fixed-block format.
3. Fixed zero (lower left corner of machine table).

4. Control reads two whole numbers and four decimal places.
5. Preparatory functions on tape, G
   a. 82—Drill cycle.
   b. 74—Mill cycle.
   c. 66—Initiate milling cut.
   d. 65—Rapid traverse to new position, spindle up.
6. Miscellaneous functions on tape, M
   a. 06—Tool change.
   b. 36—Tool change and tape rewind.
   c. 51—53—Depth cams (manually set).
   d. 08—Mist coolant on.
   e. 09—Coolant off.

For this hypothetical MCU, program two M functions in a block.

FORMAT DETAIL: FIXED BLOCK

N3, G2, x2.4, y2.4, M2*

Note in the specifications for this numerically controlled milling machine that the zero is fixed. It is necessary then that some location on the table be chosen for the workpiece. Also keep in mind that there is no provision for minus programming. The center of the milling cutter will have to be positioned in relation to the fixed zero on the machine table. This presents no difficulty, however, and it is only mentioned to keep the student aware of this consideration. The illustration will demonstrate this condition. See Fig. 5-16.

The cutter center line position A (start cut) and cutter center line position B (end cut) are shown with their respective dimensions from the zero point on the machine table in the x axis. Fig. 5-17 also demonstrates the dimensions in the y axis.

The centers of the holes are calculated in the same way, adding the distance from machine zero.

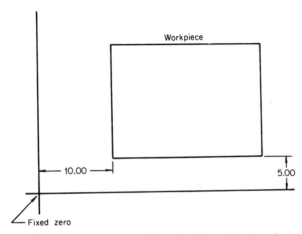

Fig. 5-16 Drawing of workpiece in Fig. 5-10 placed on machine table. Offset equals 10 inches in x axis and 5 inches in y axis from machine fixed zero.

Procedure

1. Study Figs. 5-10 and 5-16.
2. Number or letter the holes and slot.
3. Calculate coordinates for:
   a. Holes (Fig. 5-18)
   b. Slot start cut position, center of cutter $\frac{5}{16}$ inch from part
   c. Slot finish cut position (Fig. 5-17)
   d. Tool change, load-unload position
4. Do not make a new tool sheet for this exercise.
5. Prepare a tape manuscript (Fig. 5-19).
6. Prepare tapes.

In this section of the text we have covered the following:

1. Additional standardized coding for tapes and controls. However, there will be many variations on these standards.

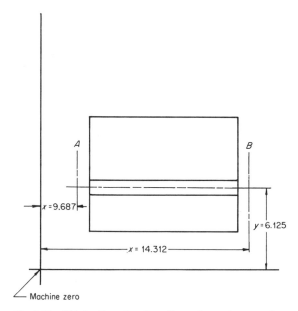

Fig. 5-17 Illustration showing dimensions of center line of milling cutter from fixed zero at start position and end position for part shown in Fig. 5-16.

2. A straight cut milling operation that was included in point-to-point programming. By straight cut is meant a cut parallel to or at a right angle to the ways of the machine. This explanation should serve to differentiate between this type of cut as compared to an angular departure.
3. Standard format detail procedure.
4. The use of a tool sheet.
5. Writing complete programs for absolute systems with these features:
   a. Full floating zero
   b. Fixed zero with a small amount of shift in x and y
   c. Tab sequential format
   d. Fixed block format

Position number	x	y
1	10.500	5.500
2	13.500	5.500
3	13.500	7.500
4	10.500	7.500
5	12.000	6.500
Tool change	17.000	11.000
A	9.6875	6.125
B	14.3125	6.125

Fig. 5-18 Coordinates for holes and milled slot, as calculated from Figs. 5-10 and 5-16.

Tape program manuscripts written for full range zero shift with the word-address format and variations of the auxiliary and miscellaneous functions should be a definite assignment at this time. Your instructor can guide you. It should also be mentioned that a great many industries are doing contouring operations on point-to-point machines and are having tremendous success. In order to accomplish this, the plants in question must have prompt access to computers for the great number of calculations required in this type of operation on point-to-point systems. See Fig. 5-20. Point-to-point systems that are used for contouring operations should be equipped with buffer storage to handle the great flow of information to the machine.

In Chap. 6 contouring or continuous-path systems will be discussed. Much of the knowledge gained from the previ-

N	G	x	y	M	E	Operator
001	82	105000	055000	51 08	E	Stop 1, C. Dr. (rpm), Feed
002	82	135000	055000	51 08	E	
003	82	135000	075000	51 08	E	
004	82	105000	075000	51 08	E	
005	82	120000	065000	51 08	E	
006	65	170000	110000	06 06	E	T. C. (rpm), Feed
007	82	105000	055000	52 08	E	Stop 2
008	82	135000	055000	52 08	E	
009	82	135000	075000	52 08	E	
010	82	105000	075000	52 08	E	
011	82	120000	065000	52 08	E	
012	65	170000	110000	06 06	E	T. C. (rpm), Feed
013	74	096875	061250	53 53	E	Stop 3, Clamp quill
014	66	143125	061250	53 08	E	Initiate milling cut
015*	74	143125	061250	53 09	E	Unclamp quill
016	65	170000	110000	36 36	E	Unload, load, T.C.

* This 74 code is used to allow the operator to unclamp the quill and retract manually.

Fig. 5-19 Program manuscript for Fig. 5-10. Fixed block format, fixed zero.

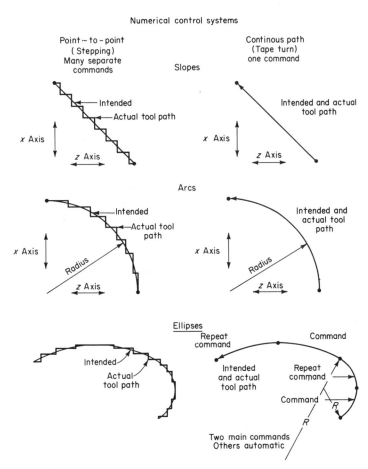

Fig. 5-20 This illustration demonstrates the differences in programming for point-to-point and continuous-path systems when angles and arcs must be cut.

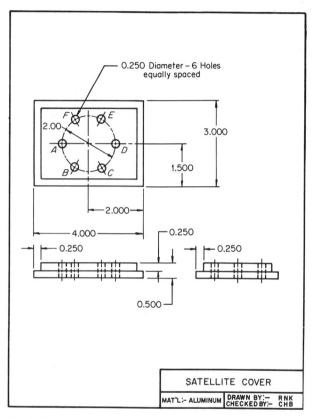

Fig. 5-21

ous chapters will expedite the learning process from now on.

Experienced machinists, technicians, and engineers should now be able to adapt to any positioning system with a minimum of effort and time. However, there is still much to learn. Only constant application and practical, actual programming work can help you to become expert in this field.

REVIEW QUESTIONS

5-1. From the standpoint of economics, why is it advantageous to have as many functions as possible controlled by tape?

5-2. What is the purpose of a preparatory function?

5-3. The table of miscellaneous functions generally describes suggested codes for what kinds of information?

5-4. Describe two methods that N/C machine manufacturers utilize to initiate a milling operation.

5-5. What is meant by a fixed or "canned" cycle?

5-6. When the milling cutter is in position and the operator clamps the quill in place, what must he then do in order to initiate the milling operation?

5-7. What does the M00 code do in a taped program?

5-8. Why do some N/C planning personnel construct mockups and visual aids for the N/C systems in the plant?

5-9. What does format detail describe?

5-10. Prepare a manuscript and tape for the workpiece in Fig. 5–21. This control shall have these features:

    a. Two axes, x and y, under tape control

    b. Absolute system

    c. Word-address format

    d. Full range zero offset; use lower left corner of workpiece as origin zero

    e. Control that reads two whole numbers and four decimal places

        1. Repeated words need not be programmed

        2. Leading zeros must be programmed

        3. Trailing zeros may be suppressed

    f. Sequence number, N, plus three digits

    g. Preparatory functions on tape, G, plus two digits

        G81—Drill cycle

        G74—Mill cycle

        G64—Initiate milling operation to end of cut

        G80—Cancel cycle

    h. Miscellaneous functions on tape, M, plus two digits

        M06—Tool change

        M36—Tool change and tape will rewind

        M51—

        M52—Manually set depth cams

        M53—

        M07—Flood coolant on

        M09—Coolant off

All other functions are manually set.

6

Continuous-Path Systems (Contouring)

The design of modern aircraft and the critical weight and shape tolerances required in their manufacture are a few of the reasons for the development of numerically controlled machine tools. Drafting machines, milling machines, welding applications, punch presses, and lathes are all suitable for these controls.

You will remember that the first successful machines were of the contouring or continuous-path variety. Contouring controls are systems in which the controlled path can result from the coordinated, simultaneous motion of two or more axes.

Fig. 6-1 Onsrud F-160 drilling, milling, and boring machine; three-axis, continuous path, numerically controlled. (*Onsrud Machine Works, Inc.*)

The same *principles* of programming that apply for point-to-point controls also apply for continuous-path systems. There is one consideration, however, which *must* be borne in mind when programming for contouring controls. Tool offset now becomes a prime factor in the programmer's calculations. Drilling operations are no problem in this respect since the center of the hole becomes the programmed dimension. The drill diameter determines the size of the hole.

In contouring operations the programmed coordinates are not the same as the contour to be machined. See Fig. 6-1.

In continuous-path programming, the path the machine tool cutter takes in moving from one position to another is

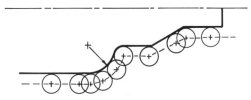

Fig. 6-2 Illustration shows path of a cutter along the profile of a workpiece with continuous-path control.

of the utmost importance. It is this path that will produce the desired contour on the workpiece. This contour may consist of straight-line cuts (including angles) and radii. Motion in one axis or more than one axis simultaneously may be required at any time in the program (Fig. 6-2).

Fig. 6-3 Sales of tape-controlled lathes have increased considerably. Photo shows a lathe final assembly section with marriage of General Electric controls to Lodge & Shipley slant-bed lathes. (*Lodge & Shipley Co.*)

Fig. 6-4   View of Lodge & Shipley slant-bed lathe, numerically controlled. Chip-removal assembly is attached.   (*Lodge & Shipley Co.*)

A "lathe contouring" system has been chosen for this section for a number of reasons. Sales of tape-controlled lathes have increased sharply. This includes engine lathes, turret lathes, and other modifications of this basic machine tool.

It really matters little whether we profile a workpiece that revolves, or perform a profiling operation on a milling machine. The principles of cutter offset programming are the same. Speeds, feeds, cutter selection, indexing of tools, and other functions are still necessary and, in many ways, very similar.

Much of the preliminary learning has already been accomplished, if point-to-point systems are thoroughly understood. Some remedial work may have to be done in this

Fig. 6-5 Shows some workpieces done on N/C lathes. (*Lodge & Shipley Co.*)

area, but in the main you should be able to progress very rapidly now.

To present this section in an orderly manner, short learning sequences will be utilized. Earlier in the text it was said that an incremental system would be used. To better understand this method of tool departures, some review will be done and some further information will be added.

INCREMENTAL DEPARTURES

An incremental departure is referenced from the present tool position. In the departure problems in Fig. 6-6 you will note that departures are not referenced to the origin for each move. It is important to understand that the absolute coordinate of a point cannot be programmed to cause a tool to be positioned at the required point. The machine control unit does not know its location in relation to absolute zero. The control knows only that it has been commanded to go a specific distance along one or more axes of motion. (See Figs. 6-6 and 6-7.)

Before you do the calculations for incremental departures, study the axis nomenclature for the numerically con-

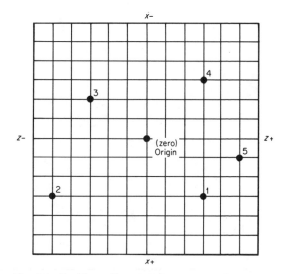

Fig. 6-6 Problem: Program incremental departures from 1 to 2, 4 to 3, 4 to 1, 4 to 2, 3 to 5, and 2 to 4.

Departures	x	z
1 to 2	$x0$	$z - 8$
4 to 3	$x + 1$	$z - 6$
4 to 1	$x + 6$	$z0$
4 to 2	$x + 6$	$z - 8$
3 to 5	$x + 3$	$z + 8$
2 to 4	$x - 6$	$z + 8$

Fig. 6-7 Solutions for incremental departure problems in Fig. 6-6.

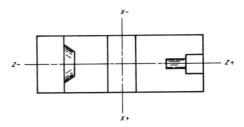

Fig. 6-8 x Minus direction toward center line of lathe decreases diameter of work. z Minus departure toward headstock. z Plus departure toward tailstock. x Plus departure out from center line of lathe increases diameter.

trolled lathe. If you are puzzled by the address character Z, a simple rule of thumb follows: Machine tool builders generally designate the spindle of a machine tool as the z axis (Fig. 6-8).

To make this section more easily understood, our own control system will be designed that will do many of the contouring operations normally included in most continuous-path MCU's.

The final goal will be to guide a cutting tool around the profile of a workpiece that will include the following types of operations:

1. Straight-line departures parallel to the x or z axes
2. Slopes or angles
3. Arcs
4. Threading

STRAIGHT-LINE DEPARTURES PARALLEL TO x AND z

Assume for this programming problem that a continuous-path control having the following features will be used:

1. Sequence (block) number readout, N plus three digits
2. Preparatory functions, G plus two digits

   NOTE: The tables of standardized codes in Chap. 5 will be used again for this system along with some additional letter address characters.

   G01—Straight-line departures (parallel to x or z).

   G04—Dwell code. Control units vary in the amount of time allowed for dwelling. For this control assume the capability of dwelling from 1 to 99 seconds. When G04 is programmed use the x distance counters to denote seconds of dwell. For example, G04 $x05$ = 5-second dwell. A dwell allows time for certain things to occur without danger to the setup or machine. Indexing of a cutting tool into position or a gear-range change of some kind should only take place when the turret or spindle is stopped to allow time for these operations to take place.

3. Distance commands: x plus five digits, z plus five digits (0.0000 to 9.9999)

   These commands will be incremental departures.

   $a.$ Minus sign must be programmed.

   $b.$ Plus sign need not be programmed.

   $c.$ Trailing zeros may be suppressed.

   $d.$ Leading zeros *must* be programmed to establish the decimal point.

4. Feed function, F plus three digits

   E.I.A. recommendations are that feed or speed "be expressed as a three (3) digit coded number." You probably will read or hear the expression "Magic 3 codes" used to describe this method of coding.

Desired speed, rpm	Code
1,728	717

The second and third digits in the code equal the desired speed (or feed) rounded to two digit accuracy (17). The first digit in the code is a decimal multiplier, and has a value three (3) greater than the *number* of digits to the *left* of the decimal point. (1,728 has four digits. Add 3 to 4 and get 7 as the first digit in the code.)

Desired feed, ipm	Code
.00875	188

If there are no digits to the left of the decimal point, the *number of zeros* immediately to the right of the decimal point are *subtracted* from 3 to provide the value of the first digit.

The second and third digits in the code equal the desired feed (speed) rounded to two-digit accuracy. The control shall have a feed range of 1 to 15 inches per minute.

For rapid traverse to positions prior to cutting, indexing, etc., the control shall have a range of 100 to 125 inches per minute.

Some controls use a feed-rate command using four digits (000.0). A feed-rate command for 5.2 ipm would be written F0052. They do not utilize any special coding for direct feed rates.

The "feed function" for combined linear and rotary motions will be explained when programming is demonstrated for angles (slopes) and arcs.

5. Miscellaneous functions, M plus two digits

M03—Spindle on, CW (See Fig. 6-9).

M05—Stop spindle

M07—Coolant on (flood)

M09—Coolant off

M40—Feed-gear range

M41—Traverse-gear range

NOTE: *a.* The control allows more than one M function to be programmed in a block if not contradictory.

b. Allow a 2-second dwell for changing gear range.

6. Other address characters, I and K plus five digits (0.0000 to 9.9999)

In Chap. 5 it was mentioned that other letter-address characters have been assigned for other functions. The E.I.A. Standards Publication RS-274-A (April, 1965) also enumerates recommendations for using the following:

I—Distance to arc center or thread lead parallel to x
J—Distance to arc center or thread lead parallel to y
K—Distance to arc center or thread lead parallel to z

For this control select the letter-address character I to be used with x, and K to be used with z.

EXPLANATION When the control receives a G01 command for a straight-line departure and senses five 9s in the I or K register, it knows that the motion is parallel to the respective x or z axis. For example, G01 x 2 I99999 equals a 2-inch departure in the plus direction, parallel with the x axis.

NOTE: As indicated, these letter-address characters have specific functions when programming arcs, angles, or threads. These other conditions will be demonstrated at the proper time.

7. Spindle speed command (rpm), S plus three digits

The Magic 3 code applies here as explained under "Feed Function." For purposes of simplicity assume only one speed for the first programming problem (Spindle speed available—500 rpm; Code—650).

NOTE: Many controls have the capability of increasing or decreasing the programmed feeds or speeds by means of dials on the control and machine.

8. Tool command, T plus two digits

Assume for this lathe that four tool positions are avail-

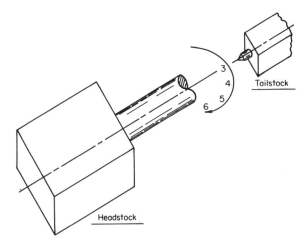

Fig. 6-9 Determine direction of spindle rotation by standing behind headstock and looking toward tailstock. M03 (clockwise rotation, forward). M04 (counterclockwise rotation, reverse).

able. If it is desired to index tool 2 into cut position, the command would be written as T22. The first digit signifies tool 2, and the second digit may reserve a tool offset position on the control. In many instances, it may be necessary for the operator to manually adjust tool position.

Programming procedure

I. Study the drawing (Fig. 6-10). Note that for this program the cutter departs from a start position at Point 1. Compensation for the radius on cutting tools is a function the programmer performs, but it will not be considered here. It is felt that programming exercises at this point can more clearly be presented if cutter radii compensation is not introduced until later on in the chapter.

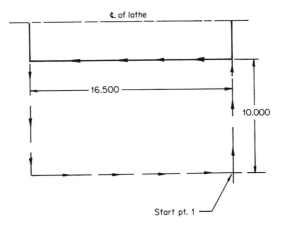

Fig. 6-10 Sample profiling operation.

II. There are numerous details to consider when starting to
plan the program manuscript for this problem. The
control unit has to be given every item of information
that is necessary to complete the task put to it. Consider the following:
  A. Starting the machine
      1. Speed selection (rpm)
      2. Direction of spindle
      3. Traverse or feed-gear range
      4. Tool index into cut position
      5. Dwell time needed to allow these functions to be
         initiated (In this hypothetical control use a 2-
         second dwell to allow any one or all of these
         functions to occur.)
  B. Positioning cutting tool
     After these preliminary steps have been programmed,
     the cutting tool is moved into position. Accomplish
     this by first traversing in a rapid traverse rate to
     within $\frac{1}{4}$ inch of the final destination, and then feed

Fig. 6-11   View of a Numeriset tool presetter.   Tool lengths may be very accurately set with these devices.   (*Lodge & Shipley Co.*)

slowly into the desired point. Study the sample manuscript to see how this is done.

C. Preparing a planning sheet

This sheet is very helpful when programming for contouring controls.

N000—Allow a 2-second dwell for

    1. Traverse-gear range

    2. Spindle on, CW

    3. Spindle speed

    4. Tool selection for turret indexing

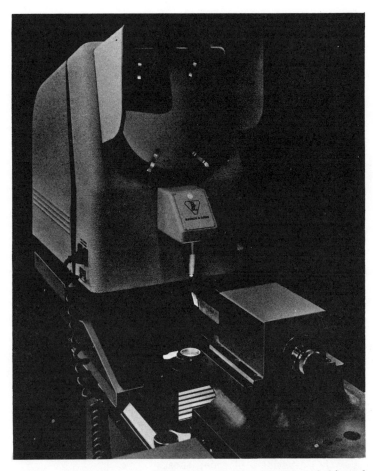

Fig. 6-12 Profile of cutter edge is shown on the comparator of this tool presetter. Backup tooling is often set up away from the machine, and need only be put on the machine tool without any adjustments being made. (*Lodge & Shipley Co.*)

N001—Command for straight-line departure to within $\frac{1}{4}$ inch of final position

N002—Allow a 2-second dwell for change to feed-gear range; coolant on

N003—Command for departure in to final-point destination

N004—Feed 9 inches toward spindle nose

N005—Feed $7\frac{1}{2}$ inches in same direction as N004

N006—Dwell 2 seconds; traverse gear range; stop spindle; coolant off

N007—Command; straight line departure out from center line; 8 inches (plus direction)

N008—Out 2 inches (plus direction)

N009—9-inch departure toward tailstock

N010—$7\frac{1}{2}$-inch departure toward tailstock (See Fig. 6-13 for manuscript.)

The person who punches the tapes for N/C programs must initiate the program by first punching an end of block code to alert the control system that a program is about to begin.

There are many interesting stories told in industrial plants of cases where an omission of this code has caused much delay and many long-distance phone calls to manufacturers of control systems.

When a program fails to start, it is wise to check this detail first before taking further steps.

LINEAR INTERPOLATION

Continuous-path control systems may be equipped with linear interpolation, circular interpolation, and parabolic interpolation.

Linear interpolation uses information in one block of information to produce velocities proportioned to the distance moved in two or more axes simultaneously.

N SEQ #	G Prep. func.	x Departure (dwell)	z Departure	I x-Arc offset (sin) (lead)	K z-arc offset (cos) (lead)	F Feed rate	S Spin. speed	T Tool select.	M Misc. func.
N000	G04	x02				F610	S650	T11	M41 M03
N001	G01	x−975		I99999					
N002	G04	x02				F510			M07 M40
N003	G01	x−025		I99999					
N004			z−9		K99999				
N005			z−75		K99999				
N006	G04	x02				F610			M05 M41 M09
N007	G01	x8		I99999					
N008		x2		I99999					
N009			z9		K99999				
N010			z75		K99999				

Fig. 6-13 Program manuscript for profiling operation in Fig. 6-10.

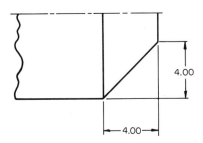

Fig. 6-14 Angular departure (x and z departures equal).

Computations are done within the control system to provide a smooth, accurate angle or taper with a minimun of programmed information. Angles and arcs will be treated separately and then be combined in a complete program manuscript.

PROGRAMMING ANGLES

In a two-axis, continuous-path control system with linear interpolation, the feed rate in each axis is controlled so that the tool is kept on the proper path. The movements must be timed to begin and end at the same time when angles are cut. Study the drawings depicting angular departures.

In Fig. 6-14 both slides have the same distance to travel, (x plus 4, z minus 4). In order to accomplish this move, the two slides must travel at the same feed rate.

In Fig. 6-15 the departure in the z minus direction must be faster than the move in the x plus direction. The ratio of z-axis departure feed to x-axis departure feed is 3 to 2.

Continuous-path controls with linear interpolation need only a minimum of programmed information to perform angle or slope operations.

To program an angle when the angle in degrees is known Program: G01 x 3464 z-6 I5 K86603 (See Fig. 6-16.)

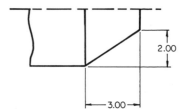

Fig. 6-15 Angular departure (*z* departure exceeds *x* departure).

NOTE: I and K values are usually written without reference to a decimal point. The control knows that I and K values (sine, cosine) are always less than one (1). It automatically shifts the decimal point in these registers when programming angles.

When the sine and cosine values are programmed, the feed rate can be provided *directly* in the desired inches per minute.

To program an angular departure when the angle in degrees is unknown but the length of the tool path is known Under these conditions when the departures are known, the length of the tool path (hypotenuse) can be closely calculated, or if the accuracy is not critical, the length of the cut can be scaled (Fig. 6-17).

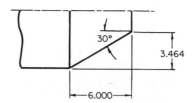

Fig. 6-16 Depicts an angle, degrees known.

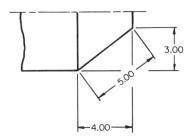

Fig. 6-17   Angle in degrees unknown;
length of tool path known.

Now, instead of providing a direct feed rate in inches per minute, it is necessary to calculate a "feed-rate number." Since no I or K values will be provided in the block of information, the control must have information that enables it to keep the tool on the desired path. A simple formula has been developed to provide this information.

$$\text{Feed-rate number} = \frac{10 \times \text{desired feed rate}}{\text{length of tool path}}$$

In Fig. 6-17 the length of the tool path is 5 inches.

$$\text{FRN} = \frac{10 \times 15}{5} = 30$$

The controls manufacturers specify a maximum of five digits to indicate feed-rate numbers (FRN = 000.00). The FRN calculated in the example would then be programmed F03 (providing trailing zeros can be suppressed). When the MCU senses no I or K values with simultaneous x and z departures, it knows that the feed function is a calculated feed-rate number and keeps the cutter on the desired path.

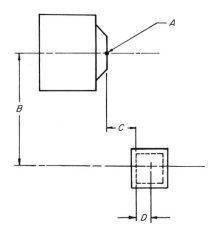

Fig. 6-18 A = location of machine zero on lathe spindle nose, B = the distance from the center of a turret to the center line of the lathe when the turret is in zero position, C = the distance from the machine zero to the surface against which a cutter is placed, D = distance from the center of the turret to the surface against which a cutter is placed.

ZERO CUT POSITION

In order to ascertain that the cutting edge of a tool is at a specific location in space it is necessary to have some reference point or zero location from which to start any program.

When programming was done for straight-line departures parallel to the x and z axes, a "start point" was established. This was done because it was convenient at that time for that specific unit lesson.

Now a new item will be introduced. Lathes may have a permanent, fixed zero location or they may have a zero shift system from which to calculate departures. In any case,

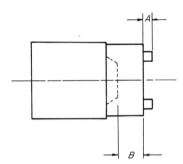

Fig. 6-19 A = width of chuck jaws, B = dimension from machine zero to face of chuck body.

the programmer must know where the tip of the cutting tool is relative to machine zero when he begins his calculations (Fig. 6-18).

In the following programming problem assume a fixed zero reference on the end of the spindle. This point now becomes the reference or origin for the first calculations to be made. Note that the dimensions are calculated from the center line of the lathe spindle for x zero position, from the left side of the turret face for z zero position. All cutting tools, workholding devices, etc., that are attached to the turret or spindle must then be considered (Fig. 6-19).

PROGRAMMING STRAIGHT-LINE DEPARTURES INCLUDING AN ANGLE

1. Study the drawing (Fig. 6-20).
2. Prepare a planning sheet (same control as previous problem).

> N000—A 2-second dwell; traverse-gear range; feed rate; tool selection; spindle on (CW) ; spindle speed

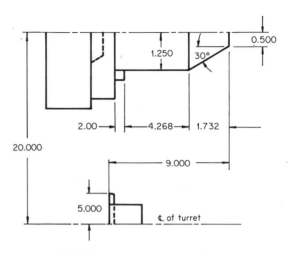

Fig. 6-20   Workpiece to be programmed.   Straight-line departures including an angle.

N001—Departure to right end of work piece (9 inches)

N002—Departure in toward edge of workpiece (9 inches)

N003—Departure in toward edge of workpiece (5 inches)

N004—A 2-second dwell; feed-gear range; feed rate; coolant on

N005—One-half inch in x minus direction to edge

N006—$x + .750$, $z - 1.732$, sin and cos 30°

N007—z-Minus direction (3.768); $\frac{1}{2}$ inch from chuck jaws

N008—Dwell; rapid traverse-gear range; feed rate; coolant off

N009—Out from center line $(x + 9)$

N010—Ditto $(x + 4.75)$

N011—z minus to home position

N012—End of program

3. Prepare manuscript (Fig. 6-21).

4. Prepare tape.

N SEQ #	G Prep. func.	x Departure (dwell)	z Departure	I x-Arc offset (sin) (lead)	K z-Arc offset (cos) (lead)	F Feed rate	S Spin. speed	T Tool select.	M Misc. func.
N000	G04	x02				F610	S650	T11	M41
									M03
N001	G01		z9		K99999				
N002		x-9		I99999					
N003		x-5		I99999					
N004	G04	x02				F510			M40
									M08
N005	G01	x-05		I99999					
N006		x075	z-1732	I5	K86603				
N007			z-3768	I99999	K99999				
N008	G04	x02				F610			M41
									M09
N009	G01	x9		I99999					
N010		x475		I99999					
N011			z-35		K99999				
N012									M02

Fig. 6-21 Program manuscript for part in Fig. 6-20.

So far in this section, then, you have

1. Calculated incremental departures
2. Identified the lathe x and z axes
3. Introduced spindle direction codes, dwell code, Magic 3 codes, I and K codes, tool selection, and linear interpolation
4. Programmed straight-line departures including an angle

It is essential at this point to understand that you have not really covered *every* facet of N/C contouring. These lessons are the blocks upon which an intelligent, curious, and ambitious programmer will build his own methods and ways of doing things.

CIRCULAR INTERPOLATION

Continuous-path systems that have circular interpolation as part of the controller package facilitate programming of circular motions. One block of information is all that is necessary to program 90° of arc or less in one quadrant. Arcs that extend into more than one quadrant require subsequent blocks for each additional quadrant.

Computations are done within the MCU to keep the cutter on the desired path.

Circular motion blocks require the proper combination of

1. Sequence number
2. Preparatory code
3. x-Axis departure
4. z-Axis departure
5. I and/or K values
6. Feed-rate number

The block may also contain

1. Optional stop
2. Speed code (rpm)
3. Coolant code

Many controls limit arc radius to 9.9999 inches.
When programming for arcs it is helpful to follow these
procedural steps:

1. Determine arc direction.
2. Determine values of x and z departures.
3. Calculate arc center offset (I and/or K).
4. Calculate a feed-rate number.
5. Program the block of information.

ARC DIRECTION

To determine whether a circular cut is in a clockwise or
counterclockwise direction, the plane of motion must be
viewed in the *negative* direction of the perpendicular axis
(E.I.A. Standards RS-274-A and RS-267). On a lathe the
y axis is the perpendicular axis. The negative direction in
this case, then, means how the curve appears as viewed from
underneath the lathe, looking up.

Some controls manufacturers specify arc direction by
looking along the y axis in the *positive* direction. In any
case, it is fairly simple to determine whether an arc is clockwise (CW) or counterclockwise (CCW). This simply points
up once again that the programmer must know the MCU
he is concerned with.

For ease of programming, then, deviate from the E.I.A.
standards for arc direction, and designate a clockwise arc
by looking along the y axis in the *positive* direction (Fig.
6-22).

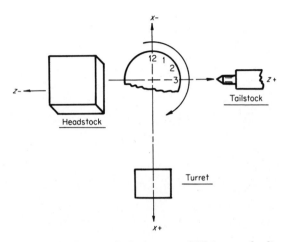

Fig. 6-22 Counterclockwise arc (E.I.A. standard). Many controls manufacturers assign a clockwise motion to this direction.

The preparatory codes for arc direction are

G02—clockwise direction
G03—counterclockwise direction

x AND z DEPARTURES FOR ARCS

Four representative arc problems are presented in Fig. 6-24. Note that these circular motions include 90° of arc in a quadrant.

CALCULATING ARC CENTER OFFSET (90° OF ARC)

In addition to using the I and K codes for angles (sine, cosine), they are also used to determine the amount of offset from the center of rotation for the start of the circular motion. The I code is utilized for the x-axis offset. The K code is used for the z-axis offset. The following rules should help when calculating I and K values for arc cutting:

Fig. 6-23 Workpiece on N/C lathe showing angle and arc cutting. Note vertical 6 position turret. (*Lodge & Shipley Co.*)

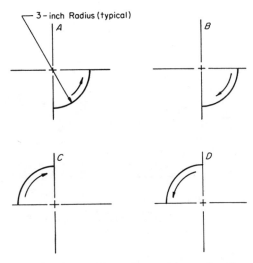

Fig. 6-24 Arc problems: Determine arc direction, x and z departures, I and K values.   The solution for Problem A is given in the text.

1. The value of I is equal to the difference in measurement between the center of rotation and the *start* of the arc in the x axial direction.

2. The value of K is equal to the difference in measurement between the center of rotation and the *start* of the arc in the z axial direction.

The "end point" of the circular motion is not considered when calculating for I and K codes. Study the 90° of arc problems in Fig. 6-24, and note that in each 90° arc either I or K will equal zero. Apply the rules for I and K codes to see how this is accomplished.

PROBLEM A SOLUTION (FIG. 6-24)   The value of I is equal to the difference in measurement between the arc center and the start of the arc in the x-axis direction. I = 3 inches.

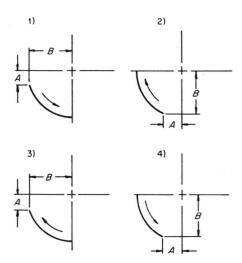

Fig. 6-25 Arc problems: Do problems 2, 3, and 4. Solution for Problem 1 is given in the text. $A = 1.250$, $B = 2.750$, radius = 3 inches (typical).

NOTE: No algebraic signs are needed for I or K values.

The value of K is equal to the difference in measurement between the arc center and the *start* of the arc in the z-axis direction. K = zero.

In Problem B the value of I is zero and K = 3 inches.

Program for 90° arc, Problem A: G03 $x - 3$ z3 I3. . .

If the value of I or K is zero, it need not be programmed.

CALCULATIONS FOR ARC CENTER OFFSET (LESS THAN 90° PER QUADRANT) (REFER TO FIG. 6-25)

Follow the steps outlined for 90° of arc and do Problems 2, 3, and 4. Solution for Problem 1 follows:

Step 1. Determine arc direction
    G03 (deviation from E.I.A. standards)

Step 2. Determine x and z departures
$\qquad x + 1.750$ inches
$\qquad z + 2.750$ inches
Step 3. Calculate arc center offset
$\qquad I = 1.250$ inches
$\qquad K = 2.750$ inches
Program G03 x175 z275 I125 K275. . .
$\qquad$ Trailing zero suppression

FEED–RATE NUMBERS

Circular cutting motions require that a calculated feed-rate number be used for advancing the cutter around the arc for some controls. Computations are done within the controller, and the desired circular path is obtained. Feed-rate numbers have been assigned three whole numbers and one decimal place (FRN = 500.0 max.). The formula for feed-rate numbers is similar to the one used for cutting angles.

$$FRN = \frac{10 \times \text{desired feed rate (ipm)}}{\text{radius of arc (inches)}}$$

PROGRAMMING STRAIGHT-LINE DEPARTURES, INCLUDING AN ANGLE AND ARC

In this programming problem a work-holding device (3-jaw universal chuck) and a vertical turret with a tool holder and cutting tool attached will be used. The problem will be to trace the outline of the workpiece with a zero radius cutter. Compensation for the radius on the cutter will not be considered at this time. The MCU will have the features given below:

Format detail: N3, G2, x+14, z+14, I14, F4, S3, T2, M2*

1. Sequence number readout, N plus three digits
2. Preparatory functions, G plus two digits
$\qquad$ G01—Straight-line departure

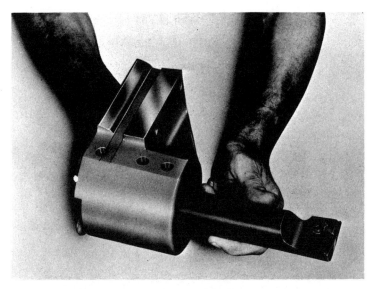

Fig. 6-26 A view of a boring bar with carbide insert. The tool holder with cutter is mounted on a vertical turret. See Fig. 6-27. (*Lodge & Shipley Co.*)

G02—CW arc; view motion in positive direction

G03—CCW arc looking down into the workpiece

G04—Dwell (time in seconds to be programmed in x register—max. = 99.9999)

3. Distance commands, terms in incremental departures

Letter address x, $z(\pm 0.0000$ to $\pm 9.9999)$

   a. Minus sign must be programmed. Plus sign need not be programmed.

   b. Trailing zero suppression.

   c. Leading zeros must be programmed to establish the decimal point.

   d. Arc center distance; letter address I, K.

Fig. 6-27 A six-position vertical turret with tool holders and cutters mounted. (*Lodge & Shipley Co.*)

4. Feed-rate number, letter address F000.0 to F500.0

   This number must always be calculated when using circular interpolation for this controller.

5. Spindle speeds: Magic 3 coding

   For this control assume spindle speeds from 100 to 2,000 rpm in increments of 10. Letter address S.

6. Turret index command, T plus 2 digits

   First digit = six turret positions. Second digit = four tool offsets on console (maximum offset = 0.0099 inch).

7. Miscellaneous functions

   a. M00—Program stop, trailing zeros must be programmed

   b. M03—Spindle direction clockwise (forward)

   c. M05—Spindle stop

   d. M40—Feed-gear range, Magic 3 coding (15 ipm max.)

   e. M41—Rapid traverse-gear range (100 to 125 ipm)

   f. M08—Flood coolant on

   g. M09—Coolant off

   h. M02—End of program

Programmer information

1. Information in N, G, F, S, and T will not transfer out unless new information is read in.

2. More than one miscellaneous function can be programmed in a block if not contradictory.

3. It is necessary to program a 2-second dwell for

   a. Gear-range changes

   b. tool-index commands

Procedure

1. Study the drawing (Fig. 6-28).

2. Locate the point of the cutter in space from machine zero (origin).

3. Locate the right edge of the part from left tip of cutter.

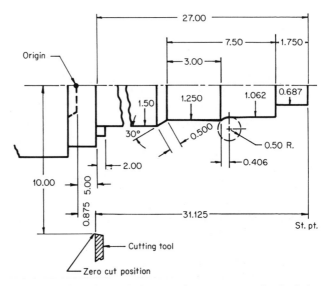

Fig. 6-28 Sample workpiece to be programmed. Includes straight-line departures with an angle and an arc.

4. Programmer instructions:
   a. 10-ipm feed rate for straight-line departures parallel to x, z.
   b. 125 ipm for rapid traverse to position.
   c. 0.3125 for approach in feed rate.
   d. Use one spindle speed (rpm)—500 for tracing outline.
   e. Trace outline to $\frac{1}{2}$ inch from chuck jaws.
   f. Return to start point (Fig. 6-29).

NOTE: In order to repeat this profiling operation with this tape as cut, it would be necessary to return to zero cut position.

Many companies are generating circular motions by calculating a series of straight-line cuts. A computer is a necessity in this case due to the great number of calculations required to approximate some arcs. Circular interpolation provides a smooth cutting path for circular movements. The requirements of individual companies will determine whether the added expense of this feature can be justified.

CUTTING THREADS ON AN N/C LATHE

A constant lead thread cut requires that the proper G, x, z, I, and K functions be programmed. The preparatory command for constant lead threads is G33.

The distance commands are: $z \pm 0$ to 9.9999 for threads parallel to the z axis. Face threads and tapered threads will not be considered here. A competent programmer should have no difficulty adapting his programs for these types of threads.

The K register serves as the lead command for z. K is programmed from 0.0001 to 0.99999 inch per thread.

Programmer information

1. When threading, an F code is not required. Any previously programmed feed codes will be ignored during the G33 mode.
2. The maximum length of thread cannot exceed 9.9999 inches per block programmed.
3. Spindle speed should be slow enough to insure electrical accuracy on some machines.
4. Lead (ipr) times spindle speed should not exceed the capacity of the clutch in active command. Feed clutch is recommended for accurate threads, but certain classes of threads can be cut in rapid traverse clutch.

N SEQ #	G Prep. func.	x Departure (dwell)	z Departure	I x-Arc offset (sin) (lead)	K z-Arc offset (cos) (lead)	F Feed rate	S Spin. speed	T Tool select.	M Misc. func.
N000	G04	x02				F612	S650	T11	M03
									M41
N001	G01		z9		K99999				
N002			z9		K99999				
N003			z9		K99999				
N004			z4125		K99999				
N005		x−9		I99999					
N006	G04	x02				F510			M40
									M08
N007	G01	x−03125		I99999					
N008			z−175		K99999				
N009		x0375		I99999					
N010			z−40937		K99999				
N011	G03	x01875	z−04062	I05	K99999	F620			

N012	G01		z−3		K99999	F510	
N013		x025	z−0433	I5	K86603		
N014			z−9		K99999		
N015			z−5817		K99999		
N016	G04	x02				F612	M05
							M41
							M09
N017	G01	x85		I99999			
N018			z9		K99999		
N019			z9		K99999		
N020			z65		K99999		
N021	G04	x02				T00	M40
N022							M02

Fig. 6-29 Program manuscript for workpiece in Fig. 6-28.

EXAMPLE   To cut 10 threads per inch

> Lead = 0.100
> rpm = 250
> 0.100 × 250 = 25 ipm
> (Clutch in active command must have this feed-rate capacity.)

CHASING A THREAD USING THE STRAIGHT-IN OR PLUNGE-CUT METHOD

PROCEDURE   See Fig. 6-30.

1. Assume that the thread diameter and thread relief are finished and the part is ready for a threading operation.
2. Coolant is on.
3. Spindle is on, 120 rpm.   (8 threads per inch = 15 ipm)
4. M40 feed-range maximum is 15 ipm.
5. PROGRAM:

	N001 G04	x02	F015	M40
1 to 2	N002 G01	x−0942	I99999	
2 to 3	N003 G33	z−66	K125	(K = lead of thread)
3 to 4	N004 G01	x0942	I99999	
4 to 1	N005 Z66	K99999		
1 to 5	N006 x−0982	I99999		
5 to 6	N007 G33	z−66	K125	
6 to 4	N008 G01	x0982	I99999	
4 to 1	N009 z66	K99999		

COMPOUND FEED COMPENSATION

The plunge-cut method usually works well when cutting fine threads such as 20, 32, 24, etc. threads per inch. However, it is normal machining practice on a conventional lathe to set the compound at one-half the thread angle to permit cutting on one side only. To accomplish this on an N/C lathe, one method is to make a z correction. See Fig. 6-31.

$$z \text{ Correction} = \text{depth of cut} \times \frac{\tan \theta}{2}$$

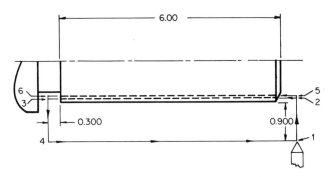

Fig. 6-30 Workpiece to be threaded.   A 60° American National Standard 8 pitch thread is to be cut.

If a 60° angle thread is being cut as in the previous program, the z correction for the second cut would be 0.023.

$$z \text{ Correction } = 0.040 \times 0.57735 = 0.023$$

The last pass should be a plunge cut to clean up both sides of the thread.

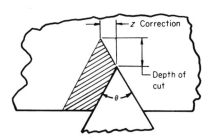

Fig. 6-31 Illustration depicting compound feed compensation. z Correction equals depth of cut times the tangent of 30°.

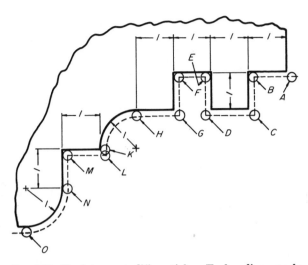

Fig. 6-32 Tool tangent differentials. Tool radius equals 0.100 inch.

In this chapter, then, manual programming for many types of cuts and operations have been discussed. Knowledge of the basic functions of programming angles and arcs should be thoroughly known and understood. Controllers that vary from any of the examples illustrated in this chapter should now be simpler to understand. The principles involved are fairly universal regardless of the manufacturer, and standards will apply in most cases.

Chapter 7 will unfold some of the aids and releases from drudgery that computer-assisted programming can provide.

TOOL TANGENT DIFFERENTIALS

A tool tangent differential is the amount that must be added or subtracted to a departure because of tool radius. Tool tangent differentials may best be explained by saying that

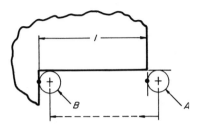

Fig. 6-33 Point A to B. Point of tangency in direction of cut did not change. Departure = 1 inch.

if the point of tangency in the direction of the cut changes, a compensation will have to be made. Fig. 6-32 illustrates many types of tool tangent differentials. A point-by-point explanation is found in the following section.

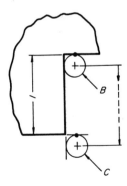

Fig. 6-34 Point B to C. Point of tangency in direction of cut did not change. Departure = 1 inch.

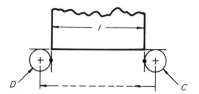

Fig. 6-35 Point *C* to *D*. Point of tangency in direction of cut changed by 180°. Departure = 1 inch plus (2 × 0.100 inch) equals 1.200 inches.

TOOL APPROACH TO TAPERED LINES

A special type of tool tangent differential is the approach of a tool to a tapered line. In any approach to this problem the programmer must use trigonometry to calculate the co-ordinates of the center of the radius of the tool as its edge contacts the tapered line. Fig. 6-44 illustrates the trigonometric analysis necessary for most tool approaches to tapered lines. See Fig. 6-45 for a sample program.

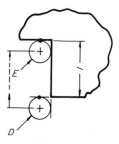

Fig. 6-36 Point *D* to *E*. Point of tangency in direction of cut did not change. Departure = 1 inch.

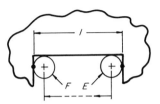

Fig. 6-37 Point E to F. Point of tangency in direction of cut changed by 180°. Departure = 1 inch minus (2 × 0.100 inch) equals 0.800 inch.

The following formulas may also be useful for certain types of tool approaches to tapered lines. In each of the sketches the radius is assumed to be known (either by direct-print dimension or by calculations).

$$A' = A \times \tan \theta$$
$$B' = B \times \sec \theta$$

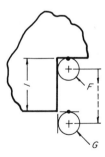

Fig. 6-38 Point F to G. Point of tangency in direction of cut did not change. Departure = 1 inch.

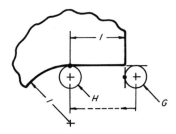

Fig. 6-39 Point G to H. Point of tangency in direction of cut changed by 90°. Departure = 1 inch plus 0.100 inch equals 1.100 inches.

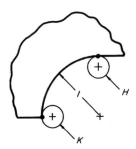

Fig. 6-40 Point H to K. Point of tangency in direction of cut changed by 90°. Departure = 1 inch minus 0.100 inch equals 0.900 inch. I equals 1 inch minus 0.100 inch equals 0.900 inch; $K = 0$.

Fig. 6-41 Point *K* to *L*. Point of tangency changed by 90°. Departure = radius = 0.100 inch. Point *L* to *M*, same as Fig. 6-33.

where *A* = distance from face of workpiece to center line of tool radius, *B* = tool radius plus amount of stock to be left on workpiece, and *x* = distance from center line of tool radius to center line of machine tool in *x* axis.

The following sketches show the angles involved in a few typical conditions of a tool approaching a tapered line during a cut. The approach to cut, in Figs. 6-46 and 6-47,

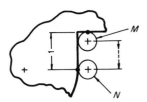

Fig. 6-42 Point *M* to *N*. Point of tangency in direction of cut changed by 90°. Departure = 1 inch minus 0.100 inch equals 0.900 inch.

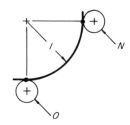

Fig. 6-43 Point N to O. Point of tangency in direction of cut changed by 90°. Departure = 1 inch plus 0.100 inch equals 1.100 inches. K = 1 inch plus 0.100 inch equals 1.100 inches; $I = 0$.

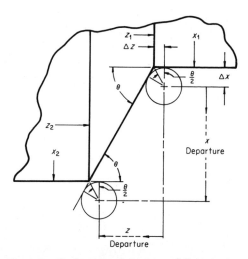

Fig. 6-44 Tool approach to tapered line. x = radius, z = radius $\times$ tan $\theta/2$, x departure = $x_2 - x_1$, z departure = x departure $\times$ ctn θ or z departure = $z_2 - z_1$.

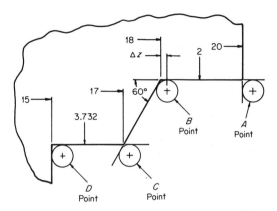

Fig. 6-45 Example of tool approach to tapered line.
A to B   G01 Z20423 K99999
B to C   X1732 Z1 I86603 K5
C to D   Z1957 K99999

is shown for reference only. To solve for x in these two sketches refer to previous examples.

The formula necessary to calculate the center point of the tool in relation to line L is given for each sketch. The formula used will also be dependent upon the direction of cut.

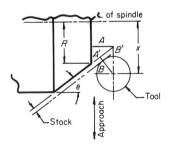

Fig. 6-46 $x = (R - A') + B'.$

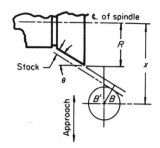

Fig. 6-47 $x = R + A' + B'$.

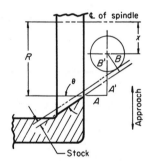

Fig. 6-48 $x = R - (A' + B')$

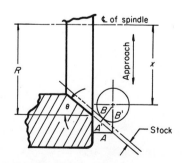

Fig. 6-49 $x = (R + A') - B'$

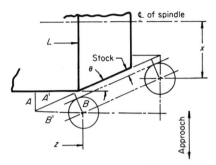

Fig. 6-50 $z = (L - A') + B'$

In each of these conditions line L is assumed to be known (either by direct-print dimensions or by calculation).

$A' = A \times \cot \theta$
$B' = B \times \operatorname{cosec} \theta$[1]

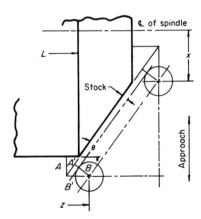

Fig. 6-51 $z = (L - A') + B'$

[1] Figures 6-32 to 6-53 are copyrighted by the Lodge & Shipley Company, and are reproduced with their permission.

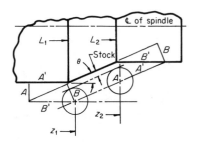

Fig. 6-52 $z_1 = (L_1 - A') + B'$
$z_2 = (L_2 + B') - A'$

where A = tool radius plus amount of stock to be left on *diameter*, B = tool radius plus amount of stock to be left on *angle*, and z = distance from zero position to center line of tool radius in z axis.

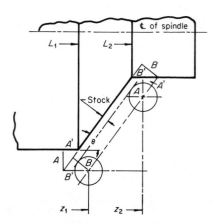

Fig. 6-53 $z_1 = (L_1 - A') + B'$
$z_2 = (L_2 + B') - A'$

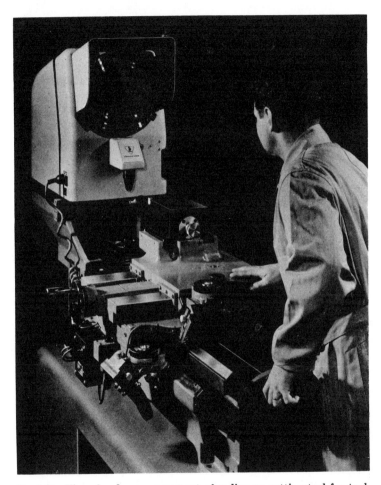

Fig. 6-54 View showing measurement of radius on cutting tool for tool tangent calculations. (*Lodge & Shipley Co.*)

REVIEW QUESTIONS

6-1. What is a contouring or continuous-path N/C system?

6-2. In a contouring system, are the programmed coordinates the same as the contour to be machined? Explain your answer.

6-3. Are the principles of cutter offset programming the same for milling machines as for lathes?

6-4. Explain what is meant by an incremental system.

6-5. What are the Magic 3 speed (rpm) codes for the following:

    a. 540 rpm

    b. 100 rpm

    c. 690 rpm

    d. 1,200 rpm

    e. 69 rpm

    f. 15 rpm

6-6. What are the Magic 3 feed (ipm) codes for:

    a. 15 ipm

    b. 125 ipm

    c. .055 ipm

6-7. Why do some machine controls provide tool offset capabilities?

6-8. Why is it necessary to depress the carriage return key before punching data for a program manuscript?

6-9. What is meant by linear interpolation?

6-10. On a lathe, x zero position is referenced from which center line?

6-11. What does "circular interpolation" do for the programmer?

6-12. How many degrees of arc may be programmed in one block of information?

6-13. What are the five steps necessary to follow when programming for arcs?

6-14. How is arc direction determined (E.I.A. Standards)? Illustrate.

6-15. How do many machine tool builders specify arc direction? Why?

6-16. Is the end point of circular motions considered when calculating I and K codes for arcs?

6-17. What do I and K codes determine when used in circular interpolation?

6-18. What is the formula for calculating feed-rate numbers for arc cutting? This formula is similar to another feed-rate number calculation. What is the other?

6-19. Is a feed code required for thread cutting on an N/C lathe? Elaborate.

6-20. For relatively coarse threads is it possible to program for thread cutting to permit cutting on only one side of the thread?

6-21. In the drawings for tool tangent differentials, substitute 0.032

for the illustrated radius in Fig. 6-32 and calculate the values of the tool departures.

6-22. Write a manuscript for tracing the outline of the workpiece in Fig. 6-55. Use the same control that was used for Fig. 6-28.

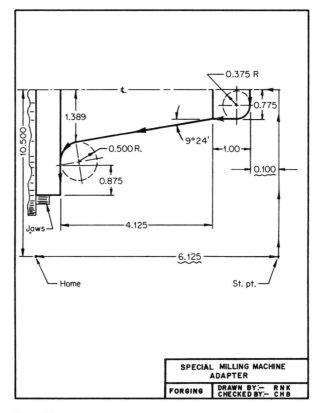

Fig. 6-55

7
N/C Programming with Computers

It is sad but true that wars, whether they be violent or simply ideological, are responsible for some of man's greatest achievements in the medical and technological fields.

World War II saw the beginning of computer applications for aircraft design and artillery-shell trajectories. Dr. Vannevar Bush at Harvard developed computers at that time.

The first electronic digital computer was a 30-ton machine developed at the University of Pennsylvania. It was called ENIAC (Electronic Numerical Integrator and Computer). This, then, was the springboard for the almost speed-of-light calculators in use today.

Electronic computers have applications in nearly every phase of business, education, traffic control, missile launching—you name the area and there will be a computer in use somewhere along the line.

The numerical control of machine tools lends itself ideally to computer-assisted programming. Tedious and time-consuming calculations can be done in microseconds on even the smallest computers. In the time it has taken to read the foregoing, a modern computer could have done literally millions of calculations.

Students, progressive manufacturers, and eager faculty in our nation's schools should have no reason to fear these electronic marvels. It is sometimes difficult to convince people that computers do not necessarily require that the user have a highly technical background in order to profit from their use. Computer manufacturers are constantly striving to make the use of their products easier to understand and less complex to use. Once a reluctant user is successful in applying a computer to his problems, he is generally quite thrilled and anxious to use it again.

CHARACTERISTICS OF COMPUTERS

To understand why computers are such powerful extensions of man's brain it is only necessary to examine four characteristics that are common to all of these machines: speed, accuracy, memory, and automatic operation. Without going into great depth, consider each of the characteristics of electronic computers.

Speed For those who are not familiar with computers, it may be a shock to discover that these intricate electronic machines work one step at a time. These are steps that many high-school or college students could perform with little or no difficulty. However, the computer can do a million or more steps in 1 second. Many com-

puters can add a quarter of a million 16-digit numbers
in 1 second!

Memory   Computers remember facts and instructions. Stored
information can be retrieved in split seconds. Computers never forget. Imagine being able to instantly
recall every fact you have ever read or learned!

Accuracy   Computers can solve the most difficult problems
in the longest but most accurate way. Computers do
not make errors.

Automatic   Once a computer has a valid set of instructions,
it can proceed to solve problems while the user goes
about other duties.

HOW COMPUTERS SOLVE PROBLEMS

Computers solve problems no differently than men do. In
working out solutions to problems there are five functions
performed. They are:

INPUT—Men gather facts together so that they can use them.

STORAGE—Men have these facts available for ready use.

CALCULATIONS—Men work out the necessary arithmetic
to find a solution.

CONTROL—Men do the operations in proper order—first
things first.

OUTPUT—Men arrive at an answer when the preceding steps
have been done.

HOW MEN TALK TO COMPUTERS

N/C programmers who use computers may never even have
seen one. How the computer gets results may be of no con-
cern to them. Some programmers manually prepare the
manuscript, and that may be all that they are responsible
for—that and the validity of their program, of course.

Writing computer programs for N/C requires that the

Fig. 7-1 Dura converter system. Computer output in the form of IBM punched cards is read and converted to tape. The tape is then used to operate N/C machine tools. Tape produced on the Dura Mach 10 or 1041C solid-state automatic typewriter is read on the Dura converter and produces IBM tab cards so that data may be fed to the computer for computations and subsequent output on tab cards. These cards may then be converted to tape to operate N/C machines. (*Dura Business Machines, division of Dura Corporation*)

programmer know the particular language for a specific kind of computer. This requirement is similar to the necessity for knowing the format for a particular machine control unit, as has been explained earlier in the text. These symbolic computer languages are not difficult to learn since they utilize words that, in many cases, are a form of "pidgin" English.

Regardless of the language required to "talk" to a given

computer, it will be very rigidly constructed and will not give correct answers if ambiguous terms or inaccurate directions are programmed.

COMPUTER LANGUAGES FOR N/C

There are literally dozens of computer languages now in existence that are used when programming for N/C machine tools. In order to eliminate much of the confusion from this sometimes troublesome area, a description of two commonly misunderstood words is in order. The words are "general processor" and "post processor." An explanation of these terms should help the learner to better understand this most interesting facet of N/C programming.

GENERAL PROCESSORS

This is a general purpose N/C computer program by which a programmer can write instructions in an easily learned language that can be interpreted by a specific kind of computer. At this point, the program will *not* produce an output for a specific N/C machine tool.

The special general processor language that the part programmer uses is set forth in manuals provided by computer manufacturers. These books are usually provided at no cost to the user. The machine tool builder is also a source of information about these programs.

POST PROCESSORS

These are computer programs that convert the general processor language into the format needed for a specific machine controller. For example, the word DRILL in a general processor statement may be converted to a preparatory command (G81) as output from the computer for a specific MCU.

In many cases the general processor and the post processor are simply decks of punched cards or rolls of magnetic tape that are fed into the computer by operators in a proc-

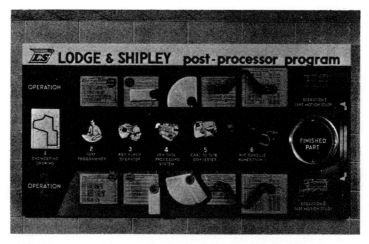

Fig. 7-2 A pictorial representation of a post-processor program. (*Lodge & Shipley Co.*)

essing center. When the part programmer delivers his manuscript to the data processing center it is usually a short wait for a tape and a printout of the taped information. Hopefully, this tape can then be put into the MCU and the workpiece machined to specifications.

The point should be made here that it is not necessary for the part programmer to know how to process his programs through a computer. It is still a good idea, however, to learn as much as possible about this area so as to make for a better and more complete understanding of the whole process.

Some general processors can be used by anyone who desires to use them. There are no legal restrictions. Other programs are proprietary and have been developed by the companies that use them. This type of program is usually restricted by the developers to "in-plant" use, and permission must be obtained in order to use them outside the plant.

Some programs require that membership fees be paid

for their use. All of these programs have had thousands of man-hours put into their development, so the user must be certain to investigate whether a program is available to him, and what restrictions, if any, there may be.

SOME POINT-TO-POINT GENERAL PROCESSORS

These programs are usually given names that indicate their use. Note the following list.

AUTOSPOT—(AUTOmatic System for POsitioning Tools).
   This program is used with the IBM 1620 and 360 computers. Post processors are needed.

AUTOPROPS—(AUTOmatic PROgram for Positioning Systems). Used with the IBM 1401 computer, this program does not require a post processor.

CAMP I—(Compiler for Automatic Machine Programming).
   Used with the General Precision, Inc., LGP 30 computer. This program was written by the Westinghouse Electric Corporation for the Milwaukee-Matic machines. It does not require a post processor.

CAMP II—This program is used with the IBM 7090 and 7094 computers. A post processor is needed.

SNAP—(Simplified Numerical Automatic Programmer).
   This is used with the IBM 1401 computer. No post processor is needed. The program was written for the Brown & Sharpe Turr-E-Tape machines.

CONTINUOUS-PATH COMPUTER PROGRAMS

There are numerous contouring general processors in use, but only a few representative programs will be discussed in this section.

APT—(Automatically Programmed Tools). This contouring program was developed in much the same way as numerically controlled machining. The Air Force sponsored a contract with the Massachusetts Institute of

Technology, and after much time and prodigious effort this processor evolved. APT also requires a post processor that produces instructions for a specific machine tool controller. It is a multiaxis program written in English-like words. In effect, the programmer figuratively rides a cutting tool along the desired path. He does this by describing the geometry of the workpiece in sections, and by determining relative positions in a cartesian coordinate system.

Very sophisticated mathematics, especially for complex curves, can be taken care of with APT. A large-scale computer must be used. Much work is still being done to improve and extend the power of this program.

AD–APT—(Air Material Command Developed APT). This is a smaller version of APT written by IBM for the Air Force. It is especially useful for programming two-dimensional configurations. AD-APT has some third-axis capability. Since it is a small version of APT, it does not need quite as powerful a computer. Post processors are usually provided by the N/C machine tool builder.

AUTOMAP—(AUTOmatic MAchining Program). This program is a two-axis contouring program suitable for programming angles and arcs with separate (not simultaneous) positioning for the third axis. It is an easy program to learn.

There are many other processors available, but all of them will not be listed here. Hopefully, a virtually universal program will soon be developed completely enough to handle nearly all N/C computer applications. In any case, there is enough similarity in these programs to choose one representative program for point-to-point and one for contouring. The examples will serve to help the learner adapt to any of the programs quite readily.

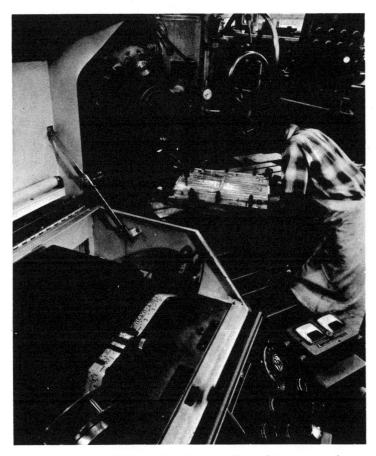

Fig. 7-3 A numerically controlled jig borer, directed by a punched tape produced by an IBM computer, automatically machines a part used in a space exploration device. The computer is able to generate the machine tool instructions in a fraction of the time required to prepare them manually. (*International Business Machines Corporation*)

It should be noted here that learning the languages of general processors and how to use them requires that the novice programmer spend many hours of study and application. While they are not usually difficult to learn, it is equivalent to a separate course of study. Textbooks and manuals must be acquired, and training should be supervised by experienced N/C programmers. Classes in computer-aided N/C programming have been set up in many factories and technical schools.

Computer manufacturers are continually modifying and updating the general processors. Newer and more powerful versions of these programs appear from time to time. Choosing a processor that will handle most of the jobs in a given shop requires careful study. If different post processors are in existence for most of the N/C machines in a plant, then more than one general processor language may have to be used.

WRITING A MANUSCRIPT USING AUTOSPOT II

AUTOSPOT I was originally written to perform point-to-point operations. The AUTOSPOT II version includes straight-line and arc milling, including tool offsets. Face- and pocket-milling instructions are also included. If no milling is required, then the second computer phase can be input directly to the post processor.

The AUTOSPOT II program consists basically of two sections, with a tool-information section added when necessary. An explanation follows.

Definition section   This part of the program defines such items as part location, remarks, tool clearance, special cutting sequences, and reference surfaces.

Machining section   This section is concerned with statements that define the operations required for machining the workpiece. Each statement in this group is structured in a specific manner, which will be explained when we write a manuscript for the computer.

Tool-information section This section describes the cutting tools required by N/C machine tools that may have taped *z*-axis control. Information such as tool numbers, taped feeds, speeds, tool diameters, tool lengths, and tool lead angles is included in this section. The format is fixed, and only necessary information is programmed.

Forms for these programs are available and quite easy to obtain from computer manufacturers and N/C machine tool builders.

It is not the intent here to teach the whole AUTOSPOT II program. Some sample applications will be shown and subsequent programs will be explained. The learner and/or his instructor should obtain as many varied programming manuals as possible in order to broaden the understanding of this very efficient aid to programmers.

A SAMPLE AUTOSPOT PROGRAM

A bolt circle has been selected to show how efficient and helpful a computer can be. Each step in the manuscript will be explained in detail. Assume that a post processor is available for the N/C machine tool. This program will be written for a three-axis Moog Hydra-Point vertical N/C milling machine with no taped speeds or feeds. These will be manually set by the operator at the machine.

The Moog Hydra-Point three-axis vertical milling machine has a fixed zero in the upper left corner of the table. An accurate subplate facilitates placing holding devices or workpieces a known distance from this fixed zero. No minus signs are needed when programming manually for this machine tool. It is necessary, however, to use minus values when using computer assists since the computer general processor does not know the features of specific machine controllers.

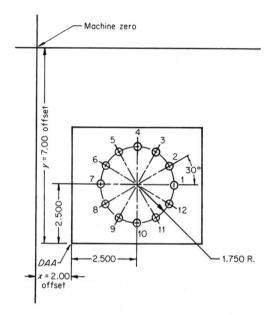

Fig. 7-4 Sample workpiece mounted on a machine table having a fixed zero in the upper left corner.

The sample workpiece has been made purposely simple to better illustrate the bolt circle routines. See Fig. 7-4.

1 REMARK/MOOG BOLT CIRCLE ROUTINE $
2 REMARK/J. DOE SEPT. 1970 $
3 DASH A (2.0, — 7.0, — 5.0) $
4 CL (0.05) $
5 SAFE (1.0, — 9.0) $
6 TOOL/DRILL 01 0.25 118.0 3.575 3.5 08 $
7 START $
8 DRILL 01/DAA, AT (2.5, 2.5) R (1.75) SA (0.0) IA (30.0) NH (12)/DP (0.5) $
9 FINI $

DESCRIPTION OF THE AUTOSPOT BOLT CIRCLE ROUTINE

1, 2. REMARK cards can be used at any time in a program. They will not appear on the tape for machining the workpiece.

3. DASH A followed by coordinate information now becomes the zero reference for the computer.

4. CL establishes the amount of clearance needed to start the cutting tool in feed rate. The cutter will rapid down to this distance from the workpiece surface and then approach in feed rate for the cutting operation.

5. SAFE plus coordinates is a selected tool change and work load-unload position that has been inserted in the post-processor deck and cannot be changed for subsequent jobs unless the post processor is changed.

6. TOOL/DRILL 01 0.25 118.0 3.575 3.5 08 $
   This tool card describes the cutter to be used. The tool number (01), diameter (0.25), lead angle (118.0), tool length (3.575), effective length (3.5), and coolant code (08) give the computer the information it needs in order to calculate z-axis departures. Effective length is equal to the tool length minus the lead.

7. START prepares the computer to accept instructions for machining the workpiece.

8. DRILL 01/DAA, AT (2.5, 2.5) R (1.75) SA (0.0) IA(30.0) NH (12)/DP (0.5) $
   This statement instructs the computer that a bolt circle is to be drilled using tool number (01) with the center of the circle at $x = 2.5$, $y = 2.5$ referenced to DASH A. The radius (R), the starting angle (SA), the incremental angle between holes (30.0), the number of holes to be drilled, and depth (DP) are all the information the computer needs to calculate the hole locations' and drilling depth.

9. FINI tells the computer that the program is ended. The dollar sign ($) which appears at the end of each

statement is equivalent to a period at the end of a declarative sentence. This sign must be used in AUTO-SPOT II.

It should be noted here that manuals for these computer programs can be procured by educators and students simply by writing to the manufacturers and indicating to them the use to which these books will be put. Every company, without exception, is interested in people who wish to know about their products and services. Another version of the AUTO-SPOT program is also available. It is called AUTOSPOT III.

AUTOSPOT II INCREMENTAL PROGRAMMING

Computers are also very helpful when programming points that are equally spaced from each other. If the locations are parallel to an axis, the technique is really quite simple to program. There are many routines available in computer general processors, and the sample problems are only shown in order to create interest in computer-aided programming.

NOTE: In Fig. 7-5 there are six equally spaced holes to be drilled parallel with the x axis. Program as follows, using a hypothetical two-axis N/C machine tool having a fixed zero on the lower left corner of the table.

1   REMARK/INCREMENTAL TECHNIQUE $
2   DASH A (4.0, 6.0) $
3   CL (0.1) $
4   SAFE (2.0, 12.0) $
5   START $
6   DRILL 01/DAA, SX(1.0) SY (2.0) EX (6.0) NH (6) $
7   FINI $

DESCRIPTION OF MACHINING STATEMENT 6 This statement alerts the computer that there are six equally spaced holes to be drilled parallel to the x axis. The computer will gen-

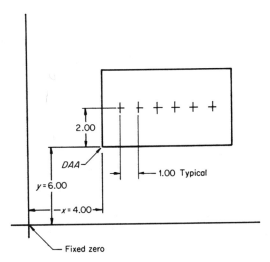

Fig. 7-5 Sample workpiece for incremental programming in AUTOSPOT II. Part is offset 4 inches in x and 6 inches in y from the machine fixed zero.

erate the necessary locations. An explanation is as follows. SX locates the center of hole 1 in the x axis referenced from DASH A (x starting point). SY locates the center of hole 1 in the y axis (y starting point). EX locates the center of hole 6 in the x axis (x end point).

There are numerous techniques and routines for programming with computer aids. Send for the manuals and master one program at a time. Keep up to date and you will have few problems.

CONTOUR PROGRAMMING WITH APT III

The APT programming system is presently the best and most thorough N/C program in existence. Extremely complex configurations can be described with its vocabulary of over 300

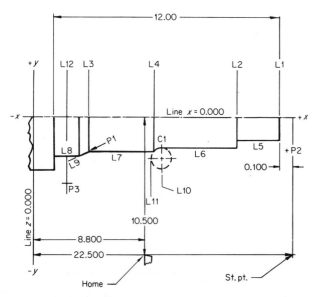

Fig. 7-6 Workpiece to be programmed in APT III. Diameters, arc, and angle are the same as the sample part in Chap. 6 (Fig. 6-28).

words. Programmers who use the APT system follow these general guidelines.

1. Describe the geometry of the part by defining all necessary points, lines, and surfaces. Since there are many ways to define arcs, circles, and lines, programs will vary depending on the individual who programs the part.
2. Tool information and cutter-path motions are described along with some functions such as feed rates, coolants, etc.

Sometimes a plot of the program is output if the necessary equipment is available. A profile that includes an angle and an arc will be programmed. A plot or drawing of the contour will also be output. This part also appeared in Chap. 6.

HOW TO WRITE A PROGRAM IN APT III

The procedures for programming the profile of the workpiece in Fig. 7-6 are as follows:

1. Identify each point, line, or circle that is needed to describe the geometry of the part by giving to each a symbol or name. The circular arc may be called Cl in the program. Lines may be identified by L1, L2, etc. Points are usually described as P1, P2, etc. The drawing of the part when labeled in this manner becomes the roadmap for the programmer.
2. Write the program.
   a. Identify the part, machine, controller, post processor used, and any other information desired. Such information might include cutter location printouts to be computed for the solution.
   b. Detail information about the labeled lines, points, circles, and surfaces. Dimensioning from the setup point on the workpiece or other locations is usual practice. The information given at this point is not necessarily dimensional. Many of the definitions simply describe the geometrical locations as they relate to other previously defined items.
   c. Tool information may then be given. This may include cutter radius, diameter, etc. Since a workpiece is being profiled with a 0.010 radius cutter this information, plus its location as loaded in the turret is also enumerated.
   d. From machine zero the cutter path is now described, with spindle commands, feed rates, coolants, etc., also listed.

The following program was written for a plotter, which accounts for words in the program such as PNDWN (pen down), CAMERA (store information), etc. The machine tool used for this program was a Jones and Lamson turret lathe with a General Electric control.

A SAMPLE APT III PROGRAM

For plotting purposes this program is set up in the cartesian coordinate system. In order to indicate z departures, line z is the y axis. Line x is the x axis. A plotter will output a drawing of the contour exactly as programmed. A tape for the specific machine tool controller is a by-product of the program.

```
 1   $$ J AND L TEST
 2              MACHIN/GECENT,1,OPTAB,77,0
 3              TURRET/FRONT,1,-.5,1
 4              CLPRNT
 5   LNX        =LINE/XAXIS
 6   LNZ        =LINE/YAXIS
 7   HOME       =POINT/8.8, -10.5
 8   RAPT       =POINT/8.7, -10.4
 9   STP        =POINT/22.5,-8.8
10   L1         =LINE/PARLEL,LNZ,XLARGE,22.4
11   L2         =LINE/PARLEL,L1,XSMALL,1.75
12   L3         =LINE/PARLEL,L2,XSMALL,7.5
13   L4         =LINE/PARLEL,L3,XLARGE,3
14   L5         =LINE/PARLEL,LNX,YSMALL,.6875
15   L6         =LINE/PARLEL,LNX,YSMALL,1.0625
16   L7         =LINE/PARLEL,LNX,YSMALL,1.250
17   L8         =LINE/PARLEL,LNX,YSMALL,1.500
18   P1         =POINT/INTOF,L3,L7
19   L9         =LINE/P1,ATANGL,30
20   L10        =LINE/PARLEL,L4,XLARGE,.406
21   L11        =LINE/PARLEL,L10,XSMALL,.5
```

22	C1	= CIRCLE/YSMALL,L6,XLARGE,L11,
		RADIUS,.5
23	L12	= LINE/PARLEL,L1,XSMALL,11
24	P2	= POINT/22.6,-1
25	P3	= POINT/11.5,-5
26		CAMERA/1
27	PARTNO	SAMPLE NO.1
28		PLOT/ALL,XYPLAN,0,-2,0,25,0,0,0
29		PENDWN
30		LEADER/20
31		TMARK/1
32		CUTTER/.010
33		TURRET/1,0,FRONT
34		FROM/HOME
35		SPINDL/1010,RPM,RANGE,LOW
36		RAPID,GOTO/STP
37		RAPID,GOTO/P2
38	INSERT	N111T01M08$
39		INDIRV/0,1,0
40		FEDRAT/10,IPM
41		G0/L5
42		TLLFT,GOLFT/L5
43		GOLFT/L2
44		GORGT/L6
45		GOFWD/C1,PAST,L7
46		GORGT/L7
47		GOLFT/L9
48		GORGT/L8,TO,L12
49		RAPID,GOTO/P3
50	INSERT	N222T09M09$
51		RAPID,GOTO/RAPT
52		FEDRAT/10,IPM
53		SPINDL/OFF
54		GOTO/HOME
55		PENUP

56	REWIND/1
57	LEADER/20
58	FINI

EXPLANATION OF THE APT III PROGRAM

Refer to the program manuscript. Each statement will not be explained in detail. Sufficient information will be given to enable the learner to understand in a general way how APT III works.

STATEMENTS 1-4   Identify the part, machine tool, post processor, tape format, turret to be used, and the number of the turret station with tool coordinates from the home position.

STATEMENTS 5-25   Identify the axes, lines, points, start point, home position, and circle that are labeled on the drawing. Some explanations follow:

8—A point just before home position to change from rapid to feed rate to offset machine inertia.
10—This statement tells the computer that line 1 is parallel to line Z and is 22.4 inches in the x plus direction (XLARGE) from line Z (0.000) as labeled on the drawing.
11—A line that is parallel to line 1 in the x minus direction (XSMALL) and 1.750 inches from line 1.
18—A point generated by the intersection of two previously defined lines (L3,L7).
22—A circle with a radius of 0.5 that is tangent to two lines (YSMALL of L6 and XLARGE of L11).

STATEMENTS 26-30   These are statements that tell the computer to store information for the plotter, initiate plot, set up parameters for plotting, bring pen down, and allow 20 inches of leader on tape.

STATEMENTS 31-33 Instructions are given to the computer to insert block number; tool information is given (radius of tip) ; and front turret position 1 is called for.

STATEMENTS 34-41 These are motion, coolant, and spindle commands that move the tool into the work from home.

STATEMENTS 42-54 These are motion commands that tell the cutter to go along surfaces as defined in the geometry section.

STATEMENTS 55-58 Plotter retracts pen. An M code is generated that will rewind tape automatically to leader. End of program.

WHEN TO USE COMPUTER AIDS

Computer-aided programming for N/C has definitely been established as feasible and economical in many manufacturing concerns. Decisions as to whether or not to utilize these calculators are influenced by many variables. A rough rule of thumb established by one programmer states that if a job is estimated to need over 3 hours of manual programming, then a computer should be used. This rule, of course, is not steadfast and probably could not stand intense scrutiny. However, it serves to illustrate one consideration, at least as to when to use a computer. There are those who advocate use of the computer for all jobs regardless of the complexity.

Send for as many varied manuals as possible. Add them to your professional library and keep on top of the latest developments by plant visits and constant reading of technical journals such as *American Machinist* and others.

REVIEW QUESTIONS

7-1. Does your organization use electronic computers for any applications? What are they used for?

7-2. What are four characteristics of computers?

7-3. How do computers solve problems?

7-4. Does a programmer for N/C have to know how to physically operate a computer?

7-5. Is it necessary to know the particular language for specific kinds of computers in order to write N/C computer programs?

7-6. Are there many computer languages for N/C? Describe one point-to-point and one contouring processor.

7-7. What is a general processor? A post processor?

7-8. Name a general processor for the following:

    a. IBM 1620

    b. LGP 30

7-9. Describe the APT general processor. AD-APT.

7-10. Does the AUTOSPOT II program need a tool-information section if only the x and y axes are fully tape controlled?

7-11. In the illustrated bolt circle routine and program (Fig. 7-4), how would you change the program to drill 20 holes? 10 holes? 15 holes?

7-12. What are the general procedures when programmaing for the APT system?

Bibliography

Automation

Amber, George H., and Paul S. Amber: "Anatomy of Automation," Prentice-Hall, Inc., Englewood Cliffs, N.J., 1962.

Drehler, Carl, and Herb Lebowitz: "Automation," W. W. Norton & Company, Inc., New York, 1957.

Computers

Berkely, Edmund C.: "Giant Brains or Machines That Think," John Wiley & Sons, Inc., New York, 1949.

Berstein, Jeremy: "The Analytical Engine," Random House, Inc., New York, 1963–1964.

Fahnestock, James D.: "Computers and How They Work," Ziff-Davis Publishing, New York, 1959.

General Electric Company: "You and the Computer," 1965.

Halacy, D. S., Jr.: "Computers, The Machines We Think With," Dell Publishing Co., Inc., New York, 1962.

Lytel, Allan: "ABC'S of Computers," Howard W. Sams & Co., Inc., The Bobbs-Merrill Company, Inc., Indianapolis, Ind., 1961.

Pfeifer, John: "The Thinking Machine," J. B. Lippincott Company, Columbia Broadcasting System, Inc., Philadelphia, 1962.

Thomas, Shirley: "Computers—Their History, Present Applications, and Future," Holt, Rinehart and Winston, Inc., New York, 1965.

Numerical control

American Society of Tool and Manufacturing Engineers, Frank Wilson (ed.): "Numerical Control in Manufacturing," McGraw-Hill Book Company, New York, 1963.

Bendix Industrial Controls Division: "N/C Handbook," The Bendix Corporation, Detroit, Mich., 1969.

Childs, James J.: "Principles of Numerical Control," Industrial Press, Inc., New York, 1965.

I.I.T.R. Institute: "APT Part Programming," McGraw-Hill Book Company, New York, 1967.

Roberts, Arthur D., and Richard C. Prentice: "Programming for Numerical Control Machines," McGraw-Hill Book Company, New York, 1968.

Numbers—Mathematics

Bergami, David, et al.: "Mathematics," Time, Inc., 1963.

Clark, Frank: "Contemporary Math," Franklin Watts, Inc., New York, 1964.

Glenn, William H., and Donavan A. Johnson: "Invitation to Mathematics," Doubleday & Co., Inc., Garden City, N.Y., 1962.

Reeves, Howard: "An Introduction to the History of Mathematics," Holt, Rinehart and Winston, New York, 1953.

Glossary

Absolute accuracy: Accuracy as measured from a reference that must be specified.

Address: Means of identifying information or a location in a control system. For example, the X in the command X 12345 is an address identifying the numbers 12345 as referring to a position on the x axis.

APT: Automatically programmed tools; a computer system for automatically programming complex parts from simple English-like instructions.

Arc clockwise: An arc generated by the coordinate motion of two axes in which curvature of the path of the tool with respect to the workpiece is clockwise.

Arc counterclockwise: An arc generated by the coordinate motion of two axes in which curvature of the path of the tool with respect to the workpiece is counterclockwise.

Auxiliary function: A function of a machine other than control of the coordinates of a workpiece or tool; includes functions such as miscellaneous functions, feed, speed, tool selection, etc. *Not a preparatory function.*

Backlash: A relative movement between interacting mechanical parts resulting from looseness; lost motion.

Binary code: A code in which each allowable position has one of two possible states. A common symbolism for binary states is 0 and 1. The binary number system is one of many binary codes.

Binary coded decimal system (BCD): A system of number representation in which each decimal digit is represented by a group of binary digits forming a character.

Binary digit: Must be either zero or one. It is equivalent to an "on" or "off," or a "yes" or "no," condition.

Bit: (*a*) An abbreviation of "binary digit." (*b*) A single character of a language employing exactly two distinct kinds of characters.

Block: A word or group of words considered as a unit, and separated from other such units by an end-of-block character.

Block count readout: Display of the number of blocks that have been read from the tape; derived by counting each block as it is read.

Buffer storage: An intermediate storage medium between data input and active storage.

Channel: See Track.

Character: One of a set of elementary marks or events that may be combined to express information. For example, the characters normally used in numerical control include those representing the decimal digits 0 to 9, the letters of the alphabet, and special characters such as tab, end-of-block, and end-of-record.

Circular interpolation: A mode of contouring control that uses the information contained in a single block to produce an arc of a circle. The velocities of the axes used to generate this arc are varied by the control.

Closed-loop system: A system in which the output, or some result of the output, is fed back for comparison with the input, for the purpose of reducing the difference.

Code: A system of signals or characters, and rules for their interpretation. For punched or magnetic tape, a predetermined arrangement of possible locations of holes or magnetized areas, and rules for interpreting the various possible patterns.

Coded decimal code: The decimal number system with each decimal digit expressed by a code.

Command: An input variable such as a pulse, signal, or set of signals impressed by a means external to and independent of the automatic control system for a specific performance.

Constant cutting speed: The condition achieved by varying the speed of rotation of the workpiece relative to the tool, inversely proportional to the distance of the tool from the center of rotation.

Contour control system: A system in which the path of the cutting tool is continuously controlled by the coordinated simultaneous motion of two or more axes.

Controller: An apparatus of unitized or sectional design, through which commands are introduced and manipulated. A program controller has the following functions: data computation, encoding, storage, readout, process computation, and output.

Coordinates: The positions or relationships of points or planes. Designations of cartesian coordinates employed in N/C programming.

1. The z axis of motion is parallel to the principal spindle axis of the machine tool. Positive z is in the direction from the work-holding means toward the tool-holding means.
2. The x axis is horizontal and perpendicular to z. If z is vertical, positive x is to the right when looking from the front of the machine toward the rear.
3. The y axis is perpendicular to both x and z. Positive y is in the direction to make a right-handed set of coordinates.

Cornering: Suddenly changing the direction of tool travel from one straight path to another.

Cutter compensation: Displacement, normal to the cutter path, to adjust for the difference between actual and programmed cutter radii or diameters.

Decimal code: A code in which each allowable position has one of ten possible states.

Diagnostic routine: A preventive maintenance check of key N/C system's components by use of a special programmed tape and/or electronic trouble-shooting instruments.

Dimension, normal: Incremental dimensions whose number of digits is specified in the format classification. For example, the format classification would be 14 for a normal dimension—X.XXXX.

Dwell: A timed delay of programmed or established duration, not cyclic or sequential, i.e., not an interlock or hold.

Encoding: Translation to a coded form from an analog or other easily recognized form without significant loss of information.

End-of-block Signal: A symbol or indicator that defines the end of one block of data.

End of program: Completion of workpiece. Stops spindle, coolant, and feed after completion of all commands in the block.

Error register or counter: A device for accumulating and signaling the algebraic difference between the quantized signal representing desired machine position and the quantized signal representing the instantaneous position of the machine.

Feed-rate override: A manual function directing the control system to modify the programmed feed rate by a selected multiplier.

Fixed-block format: A format in which the number and sequence of words and characters appearing in successive blocks is constant.

Fixed sequential format: Means of identifying a word by its location in the block. Words must be presented in a specific order, and all possible words preceding the last desired word must be present in the block.

Flip-flop: A physical element or circuit having two stable static states, either of which may be induced by means of suitable input signals.

Floating zero: A characteristic of a numerical machine tool control permitting the zero reference point on an axis to be established readily at any point in the travel. The control retains no information on the location of any previously established zeros.

Format classification: A means, usually in an abbreviated notation, by which the motions, dimensional data, type of control system, number of digits, auxiliary functions, etc., for a particular system can be denoted.

Format detail: Describes specifically which words and of what length are used by a specific system in the format classification.

Format, input media: Physical arrangement of possible locations of holes or magnetized areas. Also, the general order in which information appears on the input medium.

Hold: An untimed delay in the program, terminated by an operator or interlock action.

Incremental coordinates: Coordinates measured from the preceding value in a sequence of values.

Incremental feed: A manual or automatic input of preset motion command for a machine axis.

Instruction: A word or part of a word that tells the director to cause some operation to be performed.

Interpolator: A device for defining the path to be followed, and the rate of travel of a cutting tool or of a machine-tool slide or element when supplied with a coded mathematical description of the same. It provides the link between programmed points and smooth curves.

Language: The association of symbols, or the rules used, to represent code information.

Leader: The section of tape that appears ahead of and after a section of coded tape. When producing leader by punching tape feed, sprocket holes are produced. It may also include a parity code.

Linear interpolation: A mode contouring control that uses the information contained in a block to produce velocities proportioned to the distance moved in two or more axes simultaneously.

Manual data input: A means for the manual insertion of numerical control commands.

Manuscript: An ordered list of numerical control instructions. (See Programming.)

Memory: A term referring to the equipment and media used for holding information in machining language, electrical or magnetic form.

Miscellaneous function: An on-off function of a machine, such as spindle stop, coolant on, clamp.

Noise: An unwanted stray signal in a control system, similar to radio static, that can interfere with normal operation.

Numerical control system: A system in which actions are controlled by the direct insertion of numeric data at some point. The system must automatically interpret at least some portion of this data.

Open-loop system: A control system that has no means for comparing the output with input for control purposes.

Optional stop: A miscellaneous function command similar to a program stop except that the control ignores the command unless the operator has previously pushed a button to validate the command.

Overshoot: The amount of tool overtravel that takes place when a cornering maneuver is made at high feed rates.

Parity check: Parity check is usually a hole punched in a specific level designed for verifying the count of holes for each row as it is read from the tape. Its purpose is to verify that the proper number of holes were punched and read for each row. Parity may be either all odd or all even. Incorrect reading will automatically stop the machine.

Point-to-point system: Discrete, or point-to-point, control in which the controlled motion is required only to reach a given end point, with no path control during the transition from one end point to the next.

Positioning time: The time required to rapid traverse the tool from one cutting operation to another.

Post processor: A computer routine for translating the output of a general programming routine into a machine language suitable for a specific N/C machine.

Preparatory function: A command changing the mode of operation of the control, such as from positioning to contouring, or calling for a fixed cycle of the machine.

Processor: That portion of a computer that controls the operation input and output devices and operates on the received, stored, and transmitted data. Its circuitry includes the functions of memory, logic, arithmetic, and control.

Programming: The ordered listing of a sequence of events designed to accomplish a given task.

Programming, computer: The preparation of a manuscript in computer language and format required to accomplish a given task. The necessary calculations are to be performed by the computer.

Programming, manual: The preparation of a manuscript in machine-control language and format required to accomplish a given task. The necessary calculations are to be performed manually.

Program stop: A miscellaneous function command to stop the spindle, coolant, and feed, after completion of other commands in the block. It is necessary for the operator to push a button in order to continue with the remainder of the program.

Pulse: A pattern of variation of a quantity, such as voltage or current, consisting of an abrupt change from one level to another, followed by an abrupt change to the original level.

Readout position: Display of absolute position as derived from position feed-back transducer.

Reproducibility: The ability of a system or element to maintain its output-input precision over a relatively long period of time.

Row: A path perpendicular to the edge of a tape along which information may be stored by presence or absence of holes or magnetized areas.

Sequence number: A number identifying the relative location of blocks or groups of blocks on a tape.

Sequence-number readout: Display of the sequence number punched on the tape.

Servomechanism: A power device for directly effecting machine motion. It embodies a closed-loop system in which the controlled variable is mechanical position. Usually some amplification is necessary between the relatively weak feed-back signal and the strong command signal.

Sign digit: A plus or minus sign for arbitrarily designating the positive or negative characteristic of a coordinate.

Spindle speed: The rate of rotation of the machine spindle usually expressed in terms of rpm.

Straight-cut control system: A system in which the controlled cutting action occurs only along a path parallel to linear, circular, or other machine ways.

Subroutine: A portion of the total N/C program, stored in the computer's memory, and available upon call to accomplish a particular operation, usually a mathematical calculation. At its conclusion, control reverts to the master routine.

Tab code: A space between dimensions on the tape in accordance with tab sequential format.

Tab sequential format: Means of identifying a word by the number of tab characters preceding the word in the block. The first character in each word is a tab character. Words must be presented in a specific order, but all characters in a word except the tab character may be omitted when the command represented by that word is not desired.

Tool function: A command identifying a tool and calling for its selection either automatically or manually. The actual changing of the tool may be initiated by a separate tool-change command.

Tool offset: A correction for tool position parallel to a controlled axis.

Track: A path parallel to the edge of a tape along which information may be stored by presence or absence of holes or magnetized areas.

Transducer: A device for converting one form of energy into another form, e.g., a pneumatic signal into an electric signal.

Variable block format: A format that allows the number of words in successive blocks to vary.

Verify: To check, usually automatically, one typing or recording of data against another in order to minimize human and machine errors in the punching of tape or cards.

Windup: Lost motion in a mechanical system that is proportional to the force or torque applied.

Word: An ordered set of characters that is the normal unit in which information may be stored, transmitted, or operated upon.

Word-address format: Addressing each word of a block by one or more characters that identify the meaning of the word.

Work mounting surface: That part of table or bed on which part or tool fixture is mounted, exclusive of any coolant and/or chip trough.

Zero offset: A characteristic of a numerical machine tool control permitting the zero point on an axis to be shifted readily over a specified range. The control retains information on the location of the "permanent" zero.

Zero shift: Manually orienting the slides of an N/C machine so that the zero origin of the machine is the same point as that which was programmed.

NOTE: Many of the terms in this Glossary are from the *Automation Bulletin,* vol. 3B, published by the Electronics Industries Association.

Appendix A
The Binary System of Counting

Our decimal numbering system is base 10 and positional. There is no character that has a value of 10. Any values greater than 9 take positions to the left. Positional, base-10 notation can be illustrated in the following manner: The number 278 actually contains two 100s, seven 10s, and eight 1s.

Binary counting is also positional, but since it is a base-2 system there is no single character having a value of 2. Only a 0 and 1 are needed in the binary method. Any values greater than 1 take positions to the left.

To illustrate:

Base 10		Base 2
1	=	1
2	=	10
3	=	11
4	=	100
5	=	101
6	=	110
7	=	111
8	=	1000
9	=	1001
10	=	1010

EXPLANATION Each successive position to the left of 1 has a value that is twice the preceding value.

The number 9 in binary is written 1001. This means that starting from the right there are one 1s, zero 2s, zero 4s, and one 8.

Study the table and practice writing a few binary numbers. Your instructor can guide you (Fig. A-1).

Note that values equal to one less than any positional binary value will have ones in every position to the right:

1,023 =	one	512
	one	256
	one	128
	one	64
	one	32
	one	16
	one	8
	one	4
	one	2
	one	1
	Total =	1,023

Positional binary values											Equivalent decimal number
1,024	512	256	128	64	32	16	8	4	2	1	
					1	0	0	1	0	0	36
	1	1	1	1	1	1	1	1	1	1	1,023
		1	0	0	0	1	0	1	1	0	278

Fig. A-1 Examples of binary numbers and their decimal number equivalents.

Straight binary is still being used in some applications, but the binary coded decimal (BCD) system is standard with few exceptions.   (See Glossary.)

Appendix B
An Analysis of Two Coding
Systems Devised for
N/C Tapes

ELECTRONIC INDUSTRIES ASSOCIATION

1. Odd parity—hole in track 5.
2. Coding for numbers has been explained in the text.
3. The letters of the alphabet.
   - *a.* Codes are the same for upper- and lower-case letters.
   - *b.* Letters A through I have number values of 1 through 9: A = 1, B = 2, etc. Holes are also punched in tracks 6 and 7 to designate these alphabetic characters.
   - *c.* J through R have values of 1 through 9, and an additional hole is punched in track 7.

d. S through Z have values of 2 through 9, and an additional hole is punched in track 6.

UNITED STATES OF AMERICA CODE FOR INFORMATION INTERCHANGE (U.S.A.C.I.I.)

1. Even parity—hole in track 8.
2. Upper-case characters have an additional hole punched in track 7.
3. Lower-case characters have holes punched in tracks 6 and 7.
4. The alphabet codes follow this system:
   a. Letters A through O have number values of 1 through 15. For example, the letter N will have holes punched in tracks 2, 3, and 4 which equals $8 + 4 + 2 = 14$. An additional hole is punched in track 7.
   b. Letters P through Z are equal to 0 through 10. Additional holes are punched in tracks 5 and 7.

Appendix C
Formulas for the
Programmer

1. American National Standard threads have a 60° included angle.

   $a.$ Pitch $= \dfrac{1}{\text{number of threads per inch}}$

   $b.$ Single depth $= \dfrac{0.6495}{\text{number of threads per inch}}$

   $c.$ Single depth when thread cutting tool is advanced at 30° angle $= \dfrac{0.750}{\text{number of threads per inch}}$

   $d.$ Flat $= \dfrac{\text{Pitch}}{8}$

2. Cutting speeds are calculated as surface feet per minute (fpm).

$$a.\ \text{rpm} = \frac{12\ (\text{fpm})}{\text{circumference of work or cutter}}$$

$$b.\ \text{rpm} = \frac{4\ (\text{fpm})}{\text{diameter of work or cutter}} \quad (\text{approximate})$$

$$c.\ \text{fpm} = \frac{(\text{rpm})\ (\text{circumference of work or cutter})}{12}$$

$$d.\ \text{fpm} = \frac{(\text{rpm})\ (\text{diameter})}{4} \quad (\text{approximate})$$

3. Feed formulas for lathes and drill presses when the desired feed in inches per revolution is known, and it is required to calculate the equivalent feed in inches per minute.

Equivalent feed in ipm = (rpm) (desired feed in ipr)

When the feed in ipm is known, and the planner desires to calculate the equivalent feed in ipr:

$$\text{Equivalent feed in ipr} = \frac{\text{actual feed in ipm}}{\text{rpm}}$$

4. Feed formulas for milling:
   feed in ipm = (rpm) (feed per tooth in inches) (number of teeth in cutter)
5. Length of point on 118° angle drills = 0.3 (diameter of drill) (approximate)
6. Tap drill sizes (75%) full thread = diameter of tap or screw minus pitch.
7. Tap drill size for percentage of full thread = diameter of tap or screw minus (1.299 ÷ number of threads per inch × percent desired)
8. The outside diameter of number screws and taps = the number of the screw or tap × 0.013, and add 0.060.

Example

To find the diameter of a #8-32 tap, multiply 8 × 0.013, add 0.060, and get a total of 0.164.

9. To find the side of a square when the diagonal is known:
Diagonal × 0.70711 = side of square

10. To find the diagonal of a square when the side is known:
Side × 1.4142 = diagonal

Appendix D
Sources of N/C
Machine Tools

Avey Machine Tool Co.
Covington, Kentucky 41012

Beatty Machine & Manufacturing Co.
954 150th Street
Hammond, Indiana 46325

Browne & Sharpe Manufacturing Co.
Industrial Products Division
Precision Park
North Kingston, Rhode Island 02853

Cincinnati Lathe & Tool Co.
4701 Marburg Avenue
Cincinnati, Ohio 45209

DeVlieg Machine Co.
Fair Street
Royal Oaks, Michigan 48068

Ex-Cello Corporation
945 East Sater Street
Greenville, Ohio 45331

Ex-Cello Corporation
1200 Oakman Boulevard
Detroit, Michigan 48232

Farrand Controls Inc.
99 Wall Street
Valhalla, New York 10595

Fosdick Machine Tool Co.
1638 Blue Rock Street
Cincinnati, Ohio 45223

Giddings & Lewis Machine Tool Co.
142 Doty Street
Fond du Lac, Wisconsin 54935

Heald Machine Co.
10 Bond Street
Worcester, Massachusetts 01606

I. O. Johansson Co.
7248 North St. Louis Avenue
Skokie, Illinois 60078

Jones & Lamson
Division of Waterbury Farrel
522 Clinton Street
Springfield, Vermont 05156

Kearney & Trecker Corp.
11000 Theodore Trecker Way
Milwaukee, Wisconsin 53214

Lodge & Shipley Co.
3061 Colerain Avenue
Cincinnati, Ohio 45225

Monarch Machine Tool Co.
Sydney, Ohio 45365

Moog Incorporated
Hydra-Point Division
East Aurora, New York 14052

New Britain Machine Co.
224 South Street
New Britain, Connecticut 06050

Onsrud Machine Works, Inc.
7758 North Lehigh Avenue
Niles, Illinois 60648

Pratt & Whitney Machine Tool
Division of Colt Industries
Charter Oak Boulevard
West Hartford, Conn. 06010

Superior Electric Co.
383 Middle Street
Bristol, Connecticut 06012

Wales Strippit Co.
84–86 Buell Road
Akron, New York 14001

Warner & Swasey Co.
5701 Carnegie Avenue
Cleveland, Ohio 44103

Appendix E
Some Sources of N/C Controls

Bendix Corporation
Industrial Controls Division
12843 Greenfield Road
Detroit, Michigan 48227

Bunker-Ramo Corp.
1070 East 152nd Street
Cleveland, Ohio 44110

General Electric Company
Industrial Sales Division
1 River Road
Schenectady, New York 12305

W. B. Knight Machinery Co.
3920 West Pine Boulevard
St. Louis, Missouri 63108

Pace Controls Corporation
661 Highland Avenue
Needham Heights, Massachusetts 02194

Appendix F
Sources of Numerical Control Standards

Aerospace Industries Association of America
1729 De Sales Street, N.W.
Washington, D.C. 20036

Electronic Industries Association
2001 Eye Street, N.W.
Washington, D.C. 20006

National Standards Association, Inc.
1321 14th Street, N.W.
Washington, D.C. 20005

United States of America Standards Institute
10 East 40th Street
New York, New York 10016

Appendix G
Guide to Available N/C
and Computer Films

NUMERICAL CONTROL

DO-ALL COMPANY
"New Servo-Controlled Contour-Matic Machine for Increased Profits"
16 mm, sound, 10 minutes
Contour band machine

JONES & LAMSON–A TEXTRON COMPANY
"How to Lower Your Handicap"
16 mm, sound, 28 minutes
Automatic turret lathe

KEARNEY–TRECKER CORP.

"Changing Months to Minutes"
16 mm, sound, 15 minutes
Full color
Milwaukee-Matic

"Producing For Profit"
16 mm, sound, 18 minutes
Full color
Milwaukee-Matic, Series E

"The Profit Robbers"
16 mm, sound, 15 minutes
Full color
Machine Center of Future N/C

IIT RESEARCH INSTITUTE
10 West 35th Street
Chicago, Illinois 60616
"Symbolic Control"

U. R. WESSON COMPANY
"This Carbide Age"
16 mm, sound, 28 minutes
Full color

"Tools of Abundance"
16 mm, sound, 28 minutes
Full color
Carbides and production problems

COMPUTERS

IBM CORPORATION
"The Information Machine"
16 mm, sound, 10 minutes
Development of computer from primitive man

UNIVAC DIVISION-SPERRY RAND CORP.
"An Introduction to Digital Computers"
16 mm, sound, 35 minutes
Full color
What they are, how they work
In order to keep up with the new films being processed, contact manufacturers of machines and controls for current listings.

Appendix H
Numerical Control
Literature

AM on NC
Editors of American Machinist
McGraw-Hill Publications Company
330 West 42nd Street
New York, New York 10036

A Primer on Numerical Control
The Bunker-Ramo Corporation
1070 East 152nd Street
Cleveland, Ohio 44110

APT Part Programming
APT Long-Range Program Staff
IIT Research Institute
McGraw-Hill Book Company
330 West 42nd Street
New York, New York 10036

Introduction to Numerical Control in Manufacturing
American Society of Tool and Manufacturing Engineers
20501 Ford Road
Dearborn, Michigan 48128

Let's Discuss Numerical Control
K. M. Gettelman, Associate Editor
Modern Machine Shop
431 Main Street
Cincinnati, Ohio 45202

N/C Handbook
Bendix Corporation, Industrial Controls Division
Department B
12843 Greenfield Road
Detroit, Michigan 48227

Numerical Control
Bendix Corporation, Industrial Controls Division
Customer Training Department
12843 Greenfield Road
Detroit, Michigan 48227

Numerical Control in Manufacturing
American Society of Tool and Manufacturing Engineers
Frank W. Wilson (ed.)
20501 Ford Road
Dearborn, Michigan 48128

Numerical Control Justification: A Methodology
Professor Wilbert Steffy, Richard Bawol, Louis LaChance,
David Polacsek
Industrial Development Division
Institute of Science and Technology
University of Michigan
Ann Arbor, Michigan 48105

Numerically Controlled Machine Tools
H. Clifton Morse and David M. Cox
American Data Processing, Inc.
4th Floor
Book Building
Detroit, Michigan 48226

Powers in Numbers
Coleman Engineering Company
3121 West Central Avenue
P.O. Box 1974
Santa Ana, California 20505

Principles of Numerical Control
James J. Childs
The Industrial Press
New York, New York 10013

Programming for Numerical Control Machines
Arthur D. Roberts and Richard C. Prentice
McGraw-Hill Book Company
330 West 42nd Street
New York, New York 10036

Index